高等学校规划教材·语言学

科技语篇翻译教程

主编　雷晓峰　李　静
编者　雷晓峰　李　静
　　　肖　勇　李　丹
审校　［美］Michael Wherrity
　　　杨公建

西北工业大学出版社

西安

【内容简介】 本教材分上编和下编共两部分。上编讲述科技翻译所需的基础理论知识(翻译专业知识、科技文体特征、英汉科技语篇差异等),以拓宽译者的翻译视野,提高汉英语体转换意识,夯实翻译能力,为翻译实践打下坚实的理论基础;下编涉及多话题科技语篇的翻译实践,理论联系实际,注重讲解翻译过程,揭示翻译奥秘。每章后附有与现实世界紧密相关的翻译练习题,以期达到学以致用的目的。

本教材可供翻译工作者、翻译专业学生、语言专业学生、翻译爱好者以及科研工作者使用。

图书在版编目(CIP)数据

科技语篇翻译教程/雷晓峰,李静主编. —西安:西北工业大学出版社,2020.1
高等学校规划教材. 语言学
ISBN 978-7-5612-6816-2

Ⅰ.①科… Ⅱ.①雷… ②李… Ⅲ.①科学技术-英语-翻译-高等学校-教材 Ⅳ.①G301

中国版本图书馆 CIP 数据核字(2019)第 267439 号

KEJI YUPIAN FANYI JIAOCHENG
科 技 语 篇 翻 译 教 程

责任编辑:何格夫	策划编辑:何格夫
责任校对:朱辰浩	装帧设计:李 飞

出版发行:西北工业大学出版社
通信地址:西安市友谊西路 127 号　　邮编:710072
电　　话:(029)88491757,88493844
网　　址:www.nwpup.com
印 刷 者:兴平市博闻印务有限公司
开　　本:787 mm×1 092 mm　　1/16
印　　张:17.25
字　　数:453 千字
版　　次:2020 年 1 月第 1 版　　2020 年 1 月第 1 次印刷
定　　价:58.00 元

如有印装问题请与出版社联系调换

前 言

一、编写背景

改革开放 40 年来,我国发生了翻天覆地的变化,其中科技(即科学与技术)扮演了重要的角色。科技是社会发展的强劲助推器,使我国实现了毛泽东诗词中曾畅想的"可上九天揽月,可下五洋捉鳖"[①]的梦想;从航天科技产品,到日常生活用品,科技都具有深深的影响力,不断改变几乎所有领域的面貌,不断推动社会向前发展。这就是我国倡导大力发展科学技术的原因。

科技翻译助推了我国的科技发展。自明末清初科技翻译发轫以来,外国传教士和我国无数仁人志士将西方发达技术译入我国,加上我国自主的科学探索,逐渐奠定了我国科技坚实的基础。

"科学技术是第一生产力[②]",而创造力是科学技术发展的第一要素。科技创新是我国发展的新引擎,科技创新的第一要素是科技人才。当今社会需要更多的复合型人才,而要成为"在各个方面都有一定能力,在某一个具体的方面要能出类拔萃的人"[③],外语(尤其是英语)应用能力不可或缺。外刊阅读、情报搜集、论文发表等工作都离不开翻译,科技翻译能力的大小关乎现代科技人才的成长速度和成长高度。本教材的编写初衷即在于此。

二、编写思想与内容

1. 编写思想

本教材翻译基础理论与翻译实践并重,着力挖掘科技语言的翻译规律、规范和策略,宏观与微观并举,努力探索科学理性和科学规律在英、汉两种语言和文化中的转换规律、转换方法和转换技巧,培养翻译学习者汉英语体转换能力。

2. 编写内容

本教材采用"大科技语篇观[④]"(即科技语篇是所有非文学语篇),以翻译能力发展为中心,以翻译过程为导向,由基础理论入手,分层次逐步呈现翻译基础理论和实践活动,借助多样化的语篇文本实战演练,与市场活动接轨,引导并开展多话题的翻译活动,培养全面的"翻

---

① 毛泽东《水调歌头·重上井冈山》,1965.
② 邓小平在会见捷克斯洛伐克总统胡萨克时做出的重要论断,1988.
③ 百度百科.
④ 闫文培. 实用科技英语翻译要义[M]. 北京:科学出版社,2008:8.

译意识"和实用的翻译操作能力,满足社会对科技翻译人才和科技复合型人才的需求。

本教材分上、下两编。上编理论篇,首先呈现科技翻译中的一些基本概念(如科技翻译分类、科技翻译标准、科技翻译简史、科技翻译对译者的要求、科技文本特征等),然后讲述科技翻译基础知识(如科技英汉文本常见差异、科技翻译过程、科技翻译常用方法、科技长句的翻译、科技汉语无主句的翻译、科技文本中被动语态的翻译、科技术语的构成及翻译、科技文本中专有名词的翻译和数词的翻译等);下编实践篇,涉及各种语篇翻译(如科普、学术论文摘要、旅游、新闻、医学、航空、商务、计算机与网络以及出国留学语篇的翻译等)。全书每章后都布置有与现实世界紧密相关的翻译任务供学生练习,以期达到学以致用的目的。

三、本教材特色

1. 注重英汉双向翻译能力培养

多年来,我国学者将大量的外文信息译介到国内,加深了我国对外国的了解程度,塑造了我国对外国的认知形象。但是,由于种种原因,外国对我国认知水平较低,一定程度上阻碍了我国的对外交流,所以我国政府提出"走出去"、"一带一路"建设等战略并不断推进,坚持"引进来"和"走出去"并重。对外翻译的重担主要落在我们中国学者的肩上,对外翻译能力是"讲好中国故事"的重要保证。因此,与很多翻译教材仅强调英译汉单向能力不同,本教材注重培养英汉双向翻译能力。

2. 注重语篇翻译能力培养

尽管翻译能力是一个"自下而上"(即按照语音—词汇—小句—句子—段落—篇章顺序进行)的发展过程,语篇翻译是一个"自上而下"(即按照篇章—段落—句子—小句—词汇—语音顺序)的翻译过程,有助于更快培养语篇翻译能力。在本教材中,笔者采用了大量的语篇文本,给予详细的讲解,希望手把手地教会读者翻译时所需的语篇意识和语体转换能力。

3. 采用多话题、现实感的材料

广义上的科技是一个广阔的概念,包括除文学之外几乎所有的话题。本教材精选了与我们工作和生活紧密相关的一些话题(如科普、新闻、旅游、医学、出国留学等),采用大量具有"现实感"的网上文章进行翻译讲解和翻译练习,以方便多领域从业者学习,有助于读者在工作和生活中独自开展自己的翻译活动。

4. 注重翻译过程讲解

语言是人类进行思想交流和社会交往的工具,也是其使用者民族文化的载体,可以用来传递几乎所有信息,其复杂性不言而喻。翻译不是在真空中进行。从源语到译入语,文本穿过了怎样的"黑匣子"是笔者就具体翻译对象讲解的重点之一。

四、其他

1. 适用对象

翻译工作者、翻译专业学生、语言专业学生、翻译爱好者以及科研工作者。

2. 编写分工

主编雷晓峰为西北工业大学外国语学院教师,负责本教材的策划、立项申报、框架构建、材料收集及翻译、出版协调和大部分章节的编写工作。

主编李静为西安交通大学第二附属医院（西北医院）儿科医生，负责部分材料的收集及翻译以及第5～9章和第15章的编写工作。

肖勇和李丹为西北工业大学外国语学院教师，负责第16章"航空语篇翻译"的编写工作。

Michael Wherrity是一位美国语言学家，会使用多国语言，在语言方面极有天赋，常年住在瑞典卡尔斯塔德市（Karlstad，Sweden）。他负责本教材英语译文的审校工作。他认真、负责的审校态度和严谨、细致、睿智的译文润色工作保证了本教材高品质的英语译文。

杨公建为西安外事学院教师，负责本教材英译汉的部分审校工作。

在此，笔者对以上两位审校人员表示诚挚的感谢！

本书的出版得到了西北工业大学教务处和西北工业大学出版社的大力支持和帮助。他们的支持和帮助是笔者辛勤写作的动力源泉之一。在此，笔者对他们表示万分感谢！

由于笔者水平有限，教材中难免有不足和疏漏之处，望前辈、同仁和读者不吝批评指正。

编　者

2019年8月

目 录

上编 理 论 篇

第 1 章 科技翻译概论 ·· 3
 1.1 科技翻译简史 ·· 3
 1.2 科技文本的一般特点 ······································ 6
 1.3 科技翻译对译者的要求 ··································· 10
 1.4 科技翻译标准 ··· 12
 练习题 ·· 12

第 2 章 科技英汉文本常见差异 ···································· 15
 2.1 英语重形合,汉语重意合 ································· 15
 2.2 英语多被动,汉语多主动 ································· 16
 2.3 英语多物称,汉语多人称 ································· 16
 2.4 英语为静态语言,汉语为动态语言 ····················· 17
 2.5 英语表态在先,叙述在后;汉语叙述在先,表态在后 ··· 17
 2.6 英语主语显著,汉语话题显著 ··························· 18
 练习题 ·· 19

第 3 章 科技翻译过程 ·· 22
 3.1 理解阶段 ··· 22
 3.2 转换阶段 ··· 24
 3.3 表达阶段 ··· 24
 3.4 校对阶段 ··· 25
 练习题 ·· 25

第 4 章 科技翻译常用的方法 ····································· 28
 4.1 词和一般句子的译法 ····································· 28
 4.2 科技语篇中长句的译法 ··································· 32
 练习题 ·· 36

第 5 章 科技英语专有名词及术语翻译 ··························· 38
 5.1 科技英语专有名词的译法 ································ 38

5.2　科技英语术语的译法 ··· 41
　　练习题 ·· 44

第 6 章　科技文本中数词的翻译 ·· 46
　　6.1　确数的译法 ··· 46
　　6.2　概数的译法 ··· 47
　　6.3　增减数的译法 ··· 48
　　6.4　倍数的译法 ··· 48
　　练习题 ·· 49

第 7 章　科技汉语无主句的翻译 ·· 52
　　7.1　英汉句式建构差异 ·· 52
　　7.2　科技汉语无主句的译法 ·· 53
　　练习题 ·· 55

第 8 章　科技文本中被动语态的翻译 ·· 57
　　8.1　译为主动句 ··· 57
　　8.2　译为被动句 ··· 59
　　8.3　按套路翻译 ··· 60
　　练习题 ·· 60

第 9 章　科技语篇翻译 ·· 63
　　9.1　语篇及其特征 ··· 63
　　9.2　语篇常见的布局模式 ·· 63
　　9.3　科技语篇的翻译 ··· 65
　　练习题 ·· 72

下编　实　践　篇

第 10 章　科普语篇翻译 ·· 77
　　10.1　科普语篇的语言特点 ·· 77
　　10.2　科普语篇翻译实例 ·· 77
　　练习题 ·· 86

第 11 章　学术论文摘要翻译 ·· 89
　　11.1　学术论文摘要的构成及其特点 ·· 89
　　11.2　学术论文摘要翻译实例 ·· 90
　　练习题 ·· 95

第 12 章　计算机与网络技术语篇翻译 ······································ 98
　　12.1　计算机与网络技术语篇的特点 ······································· 98
　　12.2　计算机与网络技术语篇翻译实例 ···································· 98
　　练习题 ··· 107

第 13 章　旅游语篇翻译 ······ 109
13.1　旅游语篇的分类及特点 ······ 109
13.2　旅游语篇翻译实例 ······ 111
练习题 ······ 121

第 14 章　新闻语篇翻译 ······ 124
14.1　新闻语篇的特点 ······ 124
14.2　新闻语篇翻译实例 ······ 126
练习题 ······ 134

第 15 章　医学语篇翻译 ······ 137
15.1　医学语篇的特点 ······ 137
15.2　医学语篇翻译实例 ······ 138
练习题 ······ 146

第 16 章　航空语篇翻译 ······ 149
16.1　航空语篇的分类 ······ 149
16.2　航空语篇的特点 ······ 149
16.3　航空语篇翻译实例 ······ 150
练习题 ······ 158

第 17 章　商务语篇翻译 ······ 161
17.1　商务语篇的分类 ······ 161
17.2　商务语篇的特点 ······ 161
17.3　商务语篇翻译实例 ······ 162
练习题 ······ 168

第 18 章　出国留学语篇翻译 ······ 171
18.1　书信的翻译 ······ 171
18.2　个人简历的翻译 ······ 176
18.3　个人陈述的翻译 ······ 184
练习题 ······ 190

各章练习题参考答案 ······ 193

附录 ······ 227
附录 1　翻译补充练习 ······ 227
附录 2　翻译补充练习参考答案 ······ 248

参考文献 ······ 265

上编 理论篇

第1章　科技翻译概论

科技,即科学(science)和技术(technology)。科学解决理论问题,技术解决实际问题,两者加起来是推动社会发展的强劲动力。

狭义上讲,科技往往指的是自然科学领域的科学技术内容,比如天文学、物理学、化学和生物学等。时至今日,传统科技领域已经取得了巨大的发展,科技成果的应用改善了人类所处的环境,大大方便了人类的生活,同时加深了人类对世界的认知,又不断产生新的科技,并持续发生作用,推动社会向前发展。

广义上讲,科技英语"不仅涵盖自然科学领域的科学技术内容,而且也涉及社会科学领域的人文社会学科(如哲学、政治学、经济学、法学、历史学、伦理学、教育学等),还包括商务英语、外贸英语等以及文学理论、戏剧理论等,这些亦可统称为'社科英语'。此外,还有广告、说明书、公文、契约之类的'应用英语'也可归入科技英语的范畴"(闫文培,2008)。科技英语文本可以泛化为科技文本。在本书中,笔者采用广义上的科技概念。

如果根据文体和结构特征,从文体角度划分文本类型,文本可分为文学文本和科技文本两大类(闫文培,2008)。按照文体学上文本这样的划分方法,科技翻译即非文学文本翻译。

1.1　科技翻译简史

欲看清科技翻译的全貌,有必要了解科技翻译的发展历史。

1.1.1　明末清初的科技翻译

我国的科技翻译始于明末清初。

16世纪时,葡萄牙、西班牙、荷兰等欧洲国家已进入资本主义原始积累时期,开始海外掠夺。在偷袭、掠夺、强占我国属地遇阻后,他们派来传教士,通过译介西方科学著作,叩开了实行闭关锁国政策的我国大门。

明末清初的科技翻译渠道是来华的西方传教士,知名的有70多人,代表人物有利玛窦、汤若望、罗雅谷、南怀仁等传教士;翻译的科学著作大约有120种,其中利玛窦、汤若望、罗雅谷和南怀仁四人的译著就达75部之多。这些科学著作涉及的领域有天文学、数学、物理学、机械工程学、采矿冶金、军事技术、生理学、医学和生物学等(马祖毅,1998)。

比如,天文学方面有罗雅谷译撰的《测量全义》十卷、《恒星历表》四卷等,汤若望译的《西洋新法历书》一百卷等;物理学方面有熊三拔、徐光启合译的《泰西水法》等。

这一时期,翻译的目的是以翻译科学技术著作为名,行传播基督教之实。

这一时期,传教士在主观上主要是为了传播基督教而从事科技翻译工作的,但在客观上的确通过翻译活动把西方的一些科学技术传到了我国,在技术和思想上都产生了积极而广泛的影响。

1.1.2 清末时期的科技翻译

1. 从鸦片战争到甲午战争时期的科技翻译

1840年爆发了鸦片战争,英帝国主义的坚船利炮打开了清朝从雍正元年开始实行闭关锁国的大门。自此,中国一步步变成了半殖民地、半封建社会。

魏源在其著作《海国图志》中提出"师夷长技以制夷"的著名主张。毛泽东在《论人民民主专政》中指出,"先进的中国人,经过千辛万苦,向西方国家寻求真理。"

这一时期从事翻译工作的人很多,有中国人,也有外国人,但主要是官方洋务机构(比如总理各国事务衙门、江南制造局翻译馆)和在华的教会机构;代表人物有中国的李善兰、徐寿、王季烈和马建忠等,外国的傅兰雅、罗亨利、林乐知、金楷理、伟烈亚力等。

根据江南制造局翻译馆1909年出版的《江南制造局译书提要》所录,该局共出版书籍达160种。涉及的科技领域有声学(1种)、光学(1种)、学务(2种)、天学(2种)、补遗(2种)、政治(3种)、商学(3种)、格致(3种)、地学(3种)、工程(4种)、电学(4种)、史志(6种)、船政(6种)、交涉(7种)、算学(7种)、图学(7种)、化学(8种)、农学(9种)、矿学(10种)、附刻(10种)、医学(11种)、兵制(12种)、工艺(18种)、兵学(21种)等(马祖毅,1998;范祥涛,2011)。

主要译作如下:

数学方面,比如傅兰雅、华蘅芳述的《代数术》;傅兰雅译、江衡述的《算式集要》;傅兰雅评、华蘅芳述的《三角数理》等。

物理学方面,比如傅兰雅译、徐建寅述的《电学》;傅兰雅译、王季烈述的《通物电光》;金楷理译、赵元益述的《光学》等。

化学方面,比如傅兰雅译、徐寿述的《化学鉴原》《化学考质》《化学求数》;傅兰雅译、徐建寅述的《化学分原》等。

天文学、地质学方面,比如伟烈亚力译、李善兰述、徐建寅补充的《谈天》;金楷理译、华蘅芳述的《测候丛谈》;玛高温译、华蘅芳述的《地学浅释》《金石识别》等。

医学方面,译介得比较系统,学术性强,医院乐于采用,主要译本有傅兰雅译、赵元益述的《儒门医学》《西药大成》;舒高第译、赵元益述的《内科理法》;傅兰雅译,徐寿、赵元益述的《法律医学》等(马祖毅,1998;范祥涛,2011)。

这一时期,翻译的目的是学习西方"长技",但传教士依然假翻译之途,行传教之实。洋务机构和在华的教会机构翻译的科学书籍虽不是一流著作,也不能反映西方科学发展水平,但在当时产生了积极的影响。

2. 甲午战争至"五四"运动时期的科技翻译

这一时期,国难当头,局势危急,如康有为所言——"寝于火薪之上"。1894年,中国在甲午战争中失败,日本在黄海一战击败北洋舰队;1904年,日、俄在中国东北发动战争。这时,中国的资产阶级改良派登上政治舞台,认为"要救国,只有维新;要维新,只有学外国",所以他们非常重视翻译。

翻译的代表人物有严复、傅兰雅和舒高第等人。

1897—1911年间共翻译日本书籍958种,其中科学249种,技术243种。此外,翻译医学书籍数量较大,至1904年,西医译著已达111种(范祥涛,2011)。这一时期翻译所涉及的科技领域众多,有数学、物理学、化学、光学、农学、医学、地理学、植物学和矿物学等。

代表作有严复所译的《天演论》《穆勒名学》《名学浅说》等科学名著;天文学家高鲁编译、商务印书馆1910年出版的《空中航行术》;秋瑾译有《看护学教程》,孙中山译有《红十字会救伤第一法》等(范祥涛,2011)。

这一时期,翻译的目的是为"救亡图存",翻译教科书,兴办新式教育。

1.1.3 民国时期的科技翻译

辛亥革命以后,特别是"五四"运动以后,我国科技翻译的规模更大,涉猎更广,义理更深(方梦之等,2018)。一批学成回国的专家学者,以译介国外新技术、新思想为己任,翻译并出版了西方科学技术著作,引进了科学技术上重要的发明和发现的原理和方法。

这一时期,翻译作品中科学书籍总计达2 545种之多。译作的原作者中很多都是各领域的代表人物,著作反映了西方近现代自然科学技术的最新发展。翻译所涉及的科技领域有数学、物理学、化学、生物学、地理学和农学等。

代表作有爱因斯坦的《相对论浅释》、罗素的《罗素算理哲学》、彭加勒的《科学与方法》、牛顿的《自然哲学之数学原理》、普朗克的《近代物理学中的宇宙观》、惠格纳的《大陆移动论》、法拉第的《法拉第电学实验研究》、赫尔姆霍斯的《能之不灭》、哥白尼的《宇宙之新观念》、欧斯伐的《化学原理》、法布尔的《昆虫记》,还有米·伊林的科普著作,比如《十万个为什么》《原子世界旅行记》等(范祥涛,2011)。

这一时期,翻译的目的是以翻译科学技术著作为名,行传播基督教之实。

这一时期,翻译最突出的特点是翻译对象是西方的科学名著。

1.1.4 新中国建立后的科技翻译

1. 从新中国建立到改革开放前

1949年,中华人民共和国建立,党和政府十分重视外语人才的培养和科技翻译工作,科技翻译突增猛进,尤其是20世纪50年代中苏关系进入"蜜月期"后,两国科技界人员互动频繁,我国翻译界出现了大量译自苏联的作品。但是,中苏关系破裂后,科技翻译进入冰冷期。

据统计,这个时期的科技翻译总量约为15 211种,其中70%以上都是工业技术类书籍。翻译高峰出现在1952—1960年间,9年内译出科技著作12 983种,年均1 442.6种。统计显示,1949—1954年翻译各类科技书籍1 782种,译自苏联的有1 657种,占全部译作的93%;1961—1965年翻译科技书籍2 388种,译自苏联的有1 759种,占译作总数的74%(范祥涛,2011)。

这一时期,科技翻译涉及的领域大多为工业技术类,原著多取自苏联,翻译活动为国家经济建设服务。

2. 改革开放之后

1978年,我国实行改革开放政策,科技翻译活动逐渐得到恢复。

据统计,1978—1999年期间,科技著作翻译总量在24 000种左右,平均每年出版约1 200种,大大超过了我国科技翻译史上任何时期科技著作的年均翻译量。

另外,我国翻译界通过成立权威机构,出版各行各类词典,使译名统一工作得到加强和完善,科技翻译更加系统和完善。至今,科技翻译已呈现繁荣景象,极大地丰富了我国科技知识内容,不断推动我国经济向前发展。

1.2 科技文本的一般特点

随着科学技术的发展,出现了科技文体这一新的文体形式。本课程采用"科技"的广义概念,即科技文体是除文学文体之外的文体形式,内核是自然科学领域的科学技术内容(如化工、机械、医学、计算机等),学术著作、论文、产品说明书等也属于科技文体。

与文学作品相比,科技文章有自身的特点。深入了解这些特点有助于翻译实践,也有助于中英文写作。

E. g.

A computer network or data network is a digital telecommunications network which allows nodes to share resources. In computer networks, networked computing devices exchange data with each other using a data link. The connections between nodes are established using either cable media or wireless media.

Network computer devices that originate, route and terminate the data are called network nodes. Nodes can include hosts such as personal computers, phones, servers as well as networking hardware. Two such devices can be said to be networked together when one device is able to exchange information with the other device, whether or not they have a direct connection to each other. In most cases, application-specific communications protocols are layered (i. e. carried as payload) over other more general communications protocols. This formidable collection of information technology requires skilled network management to keep it all running reliably.

Computer networks support an enormous number of applications and services such as access to the World Wide Web, digital video, digital audio, shared use of application and storage servers, printers, and fax machines, and use of email and instant messaging applications as well as many others. Computer networks differ in the transmission medium used to carry their signals, communications protocols to organize network traffic, the network's size, topology and organizational intent. The best-known computer network is the Internet.

(https://en. wikipedia. org/wiki/Computer_work)

这篇节选的短文介绍了计算机网络的定义、用途和一些术语(如"节点")及其工作原理。翻译时,一定要弄懂专业术语及涉及的专业知识,信息之间的逻辑关系,忠实、通顺地用汉语再现原文信息。

【参考译文】

计算机网络或数据网络是指允许节点共享资源的数字化通信网络。在计算机网络中,连网计算设备可通过数据链接相互交换数据。节点之间通过电缆媒介或无线媒介相互连接。

启动、发送和终止数据传输的网络计算机设备可被称为网络节点。节点包括诸如个人计算机、电话、服务器和网络硬件等形式的主机。当一台设备与另一台设备交换信息时,无论两台设备是否直接相互连接,我们均可称这两台设备已连网。在大多数情况下,特定应用通信协议是分层的(即作为载荷随附的通信协议进行),而不是其他常用的通信协议。强大的信息收集技术需要娴熟的网络管理,以确保整体可靠运行。

计算机网络可支持大量的应用程序和服务,如访问万维网、数字视频、数字音频、共享使用的应用程序和存储服务器、打印机、传真机、电子邮件、即时消息应用程序及其他诸多应用程序等。计算机网络在用于信号传输的传输媒介、用于组织网络信息流量的通信协议、网络规模、网络拓扑和网络组织意图等方面存在差异。最知名的计算机网络是互联网。

要做到译文"忠实"和"通顺",还要注意科技文体以下常见的5个特点。

1.2.1　物称主语

"物称主语",即主语为物,而非人。这是由科技文章的特点所决定的,因为科学文本多描述科学事实或科学发现,涉及科学规律等客观事物,注重事物的客观性。

E. g. Two such devices can be said to be networked together when one device is able to exchange information with the other device, whether or not they have a direct connection to each other.

尽管本句主语为物"Two such devices",主句谓语为被动语态,考虑到英汉差异(①汉语人称(句子主语为人)多;②汉语句子很少用"被"字),所以笔者将本句译为:当一台设备与另一台设备交换信息时,无论两台设备是否直接相互连接,我们均可称这两台设备已连网。

E. g. A computer network or data network is a digital telecommunications network which allows nodes to share resources. In computer networks, networked computing devices exchange data with each other using a data link. The connections between nodes are established using either cable media or wireless media.

这里原文的3句话恰巧主语都是物称。

英语句子都有主谓结构,即英语句子都有至少主语和谓语;汉语句子多呈现"话题＋评价"结构。

E. g. 食堂今天吃饺子。话题是"食堂"。

有个事我想问你。话题是"有个事"。

当原文英语的主语恰好是汉语的话题时,翻译就变得简单起来。

因此,笔者将这段翻译为:计算机网络或数据网络是指允许节点共享资源的数字化通信网络。在计算机网络中,连网计算设备可通过数据链接相互交换数据。使用电缆媒介或无线媒介将节点连接起来。

1.2.2　文体质朴

与文学作品相比,科学文本很少使用富于美学修辞手法和艺术色彩的词语来表达科学事实、科学规律、科学发现等信息,语句信息结构往往平衡、匀称,句式不烦琐,语句长而不累赘,文体质朴。

E.g. The climate phenomenon that is being blamed for floods, hurricanes and early snowstorms also deserves credit for invigorating plants and helping to control the pollutant linked to global warming, a new study shows.

这句英文汉译其实并不容易。首先要弄明白词语的意思:be blamed for(因……被责备)、deserve credit for(因……值得受到称赞)、invigorate(使精力充沛)等词的意思。其次要注意英汉差异:汉语多用主动形式表意,不可将英语被动句式"that is being blamed for"译为汉语"被"字句,否则读来不流畅。最后要明确原文信息之间的关系。

因此,笔者将本句翻译为:一项新的研究显示,造成洪水、飓风和暴风雪的气候现象也有好的一面,因为这种气候现象不但有助于植物生长,而且也有助于控制与全球变暖有关的污染物。

1.2.3 语气正式、庄重

科学研究是一项严肃的工作,来不得半点马虎。科学发现及科学进步凝聚着科学工作者的心血,科学文本中记录着科学事实、发现、规律及科学上取得的点滴进步,为以后人类继承和发展提供支持,所以科学文本语气显得正式、庄重。

E.g. 注重计算机网络攻击方式及防御技术研究,有利于优化计算机长期使用的安全性能,给予用户安全使用计算机网络科学保障。因此,需要对计算机网卡的攻击方式进行必要的分析,并重视防御技术使用,确保计算机网络长期使用的安全可靠性,为其稳定发展注入活力。

这段话表达出了对计算机网络攻击方式及防御技术研究的重要性和必要性,要将这种正式、庄重的语言风格翻译进译文中。

【参考译文】

The focus on the computer network attack mode and defense technology research is helpful to optimize the security performance of the computer's long-term use and give the scientific protection to the users' secure use of computer network. Therefore, the necessary analysis on the attack mode of computer network card shall be made, and the use of defensive technologies shall be stressed, to ensure the security and reliability of the long-term use of computer network, and inject vitality into the stable development.

1.2.4 逻辑性强

与文学作品相比,科技作品(科学专著、科学论文、科学报道、试验报告、技术规范、工程技术说明、科技文献以及科普读物等)逻辑性强,具体表现为措辞严谨、推理严密、衔接合理、表意清晰、环环相扣等。

E.g. 据统计,地球上每年约发生500多万次地震,即每天要发生上万次的地震。其中绝大多数太小或太远,以至于人们感觉不到。能造成特别严重灾害的地震大约有一两次。人们感觉不到的地震,必须用地震仪才能记录下来;不同类型的地震仪能记录不同强度、不同距离的地震。

我们知道,英语为形合语言,信息之间的逻辑关系都要明示出来;汉语是意合语言,汉语信息靠意思内部衔接,明示的逻辑手段要尽量隐藏掉,这与英语恰恰相反,翻译时要将逻辑关系转换恰当。

【参考译文】

According to the statistics, there are more than 5 million earthquakes every year, i.e., over ten thousand earthquakes per day. Among them, most are too small or too far away, so that people cannot feel them. Roughly, there are about one or two earthquakes which cause especially serious disasters. Earthquakes which people cannot feel must be recorded by seismometers; different types of seismometers can record earthquakes with different intensities and distances.

我们可以看到,第一句汉语两次使用"发生",但这并不意味着这个动态的表达法要出现在英语中,因为英语是静态语言。"能造成特别严重灾害的地震"中"地震"前信息,我们可以解读成定语从句;"人们感觉不到的地震"中"人们感觉不到",同样我们把它解读成定语从句,与"地震"形成修饰和被修饰关系。

1.2.5 专业术语多

科技文本均属于某个专业领域,而每个专业领域都有很多专业术语,所以科技文本中有很多专业术语。这些专业术语在一个领域中往往有特定的意思,翻译时要格外注意。

E.g. 急性单核细胞白血病(M5)是急性髓性白血病中较常见的类型,临床表现具有髓外浸润、高白细胞计数、完全缓解率低、无病存活时间短、预后差、死亡率高、常规化疗的总体疗效有限等特点。免疫治疗是目前最有希望彻底清除白血病微小残留病灶、延长治疗后缓解期、预防复发的疗法,其中肿瘤疫苗已在部分白血病患者中获得肯定疗效。近年来,已有数个被自身CTL所识别的肿瘤抗原基因被发现,许多由这些肿瘤抗原基因编码的HLA-I类分子限制性CTL表位也被相继证实。在这些肿瘤抗原基因中,MLAA抗原(MLAA-22,MLAA-34)基因已被证实在急性单核细胞白血病中有高表达,而在正常组织中未见表达。MLAA-22为我室首次报道,它与急性单核细胞白血病(M5)关系密切。

对我们一般人而言,我们阅读这篇汉语段落可能都很难,所以翻译就显得更难了。

这段文字是一篇医学专业文章中的1个段落信息,其中有很多专业术语,比如"髓性白血病""髓外浸润""病灶""肿瘤抗原"等。作为非医学专业人士,译者翻译时要更加细心,应用医学专业的术语和原理,用英语完整再现原文信息,符合医学英语的表达规范。这些专业术语在英语中有唯一的所指。

词汇层:单核细胞白血病 monocytic leukemia;髓性白血病 myeloid leukemia;髓外浸润 extramedullary infiltration;化疗 chemotherapy;病灶 residual disease;肿瘤疫苗 tumor vaccine;肿瘤抗原基因 tumor antigen gene;预后 prognosis。

笔译是严肃的目的语写作。无论任何时候,译者都要严肃对待翻译工作,在很强的英汉差异意识下,在词层、句子层、段落层和篇章层信息转换到位后,用自己最高的写作水平重现原文信息。唯有这样,译文才会显得字正句顺,语义饱满,表达通畅。

【参考译文】

Acute monocytic leukemia (M5) is a common type of acute myeloid leukemia. Its clinical features include extramedullary infiltration, high white blood cell count, low complete remission rate, short disease-free survival time, poor prognosis, high mortality rate, limit effect of conventional chemotherapy, and et al. Currently immunotherapy is the most hopeful treatment to remove leukemia minimal residual disease completely, prolonged remission period and prevent recurrence. Among immunotherapy, tumor vaccine has presented positive effect in some leukemia patients. In recent years, several tumor antigen genes recognized by CTL itself have been found, and lots of HLA-I restricted CTL epitopes coded by tumor antigen gene have been certified one after another. In these tumor antigen gene, MLAA (MLAA-22, MLAA-34) gene have been proved high expression in Acute monocytic leukemia, while none expression in normal organ. MLAA-22 (Gene bank number: AAQ93061.1) was first reported by our department, which had a close relationship with acute monocytic leukemia (M5).

1.3 科技翻译对译者的要求

"科学技术是第一生产力",这是邓小平同志1988年6月在全国科学大会上提出的论断。科学技术在社会发展中扮演着革命性作用。由此可见,科技翻译对译者提出了很高的要求。译者必须从自身做起,努力提高各方面的素质,才能胜任这项工作。

早在20世纪70年代,德国学者Wilss(1976)提出了由多项能力组成的翻译能力模式,认为翻译能力由源语的接受能力、译语的产出能力以及超能力(即转换能力)三个成分组成。之后,国外学者(如Delisle、Nord、Schaffner等)和国内学者(如刘宓庆、杨晓荣、苗菊等)从教学、实证和认知等视角提出了翻译多项能力构成模式(马会娟,2013)。

然而,相比之下,西班牙巴塞罗那自治大学PATCE(Process in the Acquisition of Translation Competence and Evaluation,翻译能力习得过程和评估)研究小组提出的模式是翻译研究领域"迄今为止最为复杂的能力模式"(Lesznyak Marta,2007)。该小组自2000年提出翻译能力模式以来,通过实证研究,提出了翻译能力构成新模式(见图1.1和图1.2)。该模式显示,翻译能力包括双语能力、语言外能力、翻译专业知识、专业操作能力、策略能力以及心理-生理要素共6个能力。

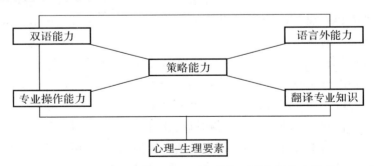

图1.1 PATCE研究小组2005年修订后的翻译能力模式

在这6个能力成分中,PATCE小组认为策略能力、专业操作能力和翻译专业知识是翻译能力所独有的。其中,策略能力在所有的能力成分中是最重要的,因为它负责解决问题并保证翻译过程富有效率,计划整个翻译项目,评价翻译过程和阶段性成果,激活其他能力以弥补不足,发现翻译问题并予以解决(马会娟,2013)。

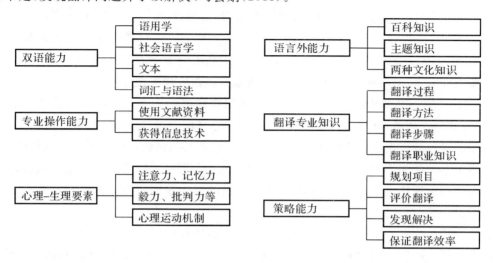

图 1.2　PATCE 小组翻译能力包含的子项目(马会娟,2013)

由此可见,该研究小组定义的"翻译能力"不但包括语言内外能力,而且包括翻译专业相关能力、生理和心理要素以及统筹规划能力,几乎涵盖了翻译实践所需的所有能力。

在 PATCE 翻译能力模式中,与科技翻译最相关的是以下三项能力。

1. 译者的双语能力

从传统意义上讲,翻译是跨语言的,通过语体转换用译入语传递出源语的信息,科技翻译也不例外。译者的双语能力主要体现在其对外语和汉语的理解与表达能力方面。具体而言,就是译者对两种语言的词汇和语法知识的掌握程度和应用能力。译者需要掌握大量的词法和科技术语,娴熟地使用语法、句法和章法,在英汉差异意识下练就强大的源语解析和译语生成能力,这样才能游刃有余地开展翻译工作。

2. 译者的翻译专业知识能力

根据 PATCE 研究小组发现,翻译专业知识具体包括翻译方法、翻译过程、翻译步骤和翻译专业知识等。这些知识为译者提供了翻译操作方法、流程和规范,为翻译工作的开展起到保驾护航的作用。在科技翻译中,信息的忠实传递是最重要的,而对于科技领域译者来说,这些翻译专业知识让译者的翻译操作不但有据可依,而且在一定程度上能更好地保证译文信息的忠实度。

3. 译者的学科专业知识能力

学科的专业知识能力指对学科专业知识的掌握程度。这一点对科技翻译至关重要。我们常说隔行如隔山,尤其是学科专业化发展的今天,学科不但继续向纵深方向发展,而且越来越呈现学科交叉趋势。任何一位译者不可能清楚地了解所有学科的知识,科技译者唯有不断学习才能较好地完成千变万化的翻译任务。

1.4 科技翻译标准

历史上,无论中国还是外国,对于翻译标准都有很多讨论,很大程度上是译者对自己翻译工作的经验总结。比如,中国历史上北朝末年及隋初,彦琮著《辨证论》(被视为我国第一部翻译专论)。他主张译经要"宁贵朴而近理,不用巧而背源"(倾向于直译),并提出"八备说"(八个条件)。古罗马著名演说家、政治家西塞罗被西方译界视作西方翻译史上的第一个翻译理论家,他提到"要成为译者的条件是翻译狄摩西尼的人必须自己也是狄摩西尼式的人物;译者应该与作者一样具有同样的演说能力"(谭载喜,2000)。

我国的科技翻译兴于明清两代,徐光启和意大利人利玛窦合作翻译了欧几里得的《几何原本》(后由李善兰和传教士伟烈亚力补充翻译完成)、《测量法义》等著作即是科技翻译的代表作。在中国近现代史上,西方发达国家科技著作的引入和翻译为我国科技发展奠定了坚实的基础。

很多翻译工作者提出了自己的翻译标准。比如,严复的"信""达"和"雅"、鲁迅的"宁信而不顺"、林语堂的"五美"(即音美、意美、神美、气美、形美)、傅雷的"神似"和钱钟书的"化境";奈达(美国)的"动态对等"、泰勒(英国)的"翻译三原则"等。

结合科技文本的性质,一般认为,科技翻译要遵循以下标准。

1. 忠实、规范

科技文本承载着科学领域的真理、事实、规律等知识。"忠实"是指翻译科技文时,译者须用符合目的语表达习惯的词法、句法、语法和章法重现原文信息;"规范"是指译文措辞符合所涉及科学领域的专业语言表达规范,包括术语使用正确,行文模式符合规范。科技译文做到准确和规范很重要,因为不准确、不规范的译文会给科学研究、生产生活等带来不良影响,甚至重大损失。

2. 通顺、易懂

任何译文都需做到"通顺",科技翻译也一样。译者须用科学、大众化的语言表达原文信息,译文读来通俗易懂,明白晓畅,文理通顺。翻译时,译者须尽可能杜绝硬译、死译现象,否则译文价值有很大损失。

3. 简洁、明晰

与"感染型"的文学作品不同,科技文属于"信息型"文本,以提供注重客观事实的信息内容为主,很少使用修辞手段,所以译文要简洁明快(尤其是英译汉时),一目了然。同时,译文须层次分明,表意清晰,这样易于读者获取科技信息。

练 习 题

一、请将下列文本译为汉语。

A Letter of Recommendation in Education

Dear Sir or Madam:

Miss Furnival Han has left me a very deep impression for her zeal and earnest in

English literature studies. I am very glad to write this letter of recommendation for her because as her teacher and also as director of her foreign studies college, I know very well her change from dislike and unfamiliarity to ardent love for English literature.

I remember she was interested in economics when she first entered university because she was highly gifted in mathematics, ranking first in college entrance examination. She was first repulsive to English literature due to lack of understanding. But after just a semester, she soon became greatly interested in the specialty. Though a first-year student, she soon acquired a penetrating comprehension and stood out as one of my best undergraduates in all my years of teaching experience.

As a second-year student, guided and encouraged by teachers returning from abroad in our department, she made up her mind to further her studies in the United States and exerted herself for this sake. She attended many professional literature forums and even observed postgraduate curricula. Apart from reading reference books with all earnest, she was well read in relevant Western professional books, immensely substantiating her knowledge. This was clearly visible from her theses and distinctively perceivable from our routine communication.

Miss Han was also a good announcer and journalist with the college broadcasting station. The programs she hosted was much to the liking of student audiences. I am proud of her performance and amazed at her literary quality. I backed her up fully when she opted for the pedagogy specialty according to her specialty advantage and personal features and the future trend of development in China. I cherish great hope that such a good student would receive better education because this would expedite the advancement of China's relevant faculty.

As a professor I greatly appreciate such an energetic and inquisitive, diligent and clever student. When she told me her desire to study abroad, I expressed full support and encouraged her to work hard for her goal. As college director, I hereby recommend her to you with self-pride and pledge that she will bring your university a new refreshing atmosphere of researches. I hope you will lend her a helping hand with her application.

(https://wenku.baidu.com/view/39f7c031ee06eff9aef80732.html?from=search)

二、请将下列文本译为英语。

联系商品销售事宜函

×××先生：

前一批货销售极佳，颇受好评，除电报中向你订购的数量外，公司同仁意欲扩大经营范围，我想请您给我寄上下述样品：

1. 6岁男式童装2套
2. 6岁女式童装2套
3. 5岁儿童套裙2套

　　并请附上有关资料与说明文字，供我们查看，宣传，倾听消费者的反映。我估计你们的产品是会受到欢迎的。一旦公司做出决定，我们将大批量向你方订货。具体事宜日后联系。请你们从速寄上样品。谢谢您对此事的关照。渴望赐复。在资料中请附上样品的价格、质量规格说明。

×××

×年×月×日

(https://wenku.baidu.com/view/8c16f39cdd88d0d232d46a00.html)

第2章　科技英汉文本常见差异

人们虽然对世界语言有不同的划分方法,但是一般划分为十大语系:①汉藏语系;②印欧语系;③阿尔泰语系;④闪含语系;⑤乌拉尔语系;⑥高加索语系;⑦南岛语系;⑧南亚语系;⑨达罗毗荼语系;⑩壮侗语系(360百科)。

英语属于印欧语系,汉语属于汉藏语系。在过去数千年中,这两种语言各自发展,至今无论在形态上还是结构上都存在巨大的差别。在翻译过程中,译者是否能做出恰当的差异转换是保证译文质量的关键。

科技英汉文本存在以下常见差异。

2.1　英语重形合,汉语重意合

形合,即用于明晰信息间的逻辑关系是显性的,在形式上清晰、严格地表现出来;意合,恰恰相反,信息间逻辑关系是隐性的,靠信息内部手段(如逻辑)衔接表意。英语和汉语分属前者和后者这两种属性语言。

E. g. Challenges being addressed by modern environmental biotechnology range from the search for microbes that will reduce acid rain by removing sulfur from power coal to the biological production of biodegradable plastics.

【解析】英语原句中信息的逻辑关系呈现清晰,如"being addressed by"(后置定语)、"that will reduce acid rain"(后置定语)、"by removing..."(方式状语)等。然而,在汉语译文中,这些明示的关系部分隐形化了。比如,"being addressed by"和被修饰词"Challenges"根据逻辑关系被译为一个单句"现代环境生物科技面临诸多挑战";定语从句"that will reduce acid rain"却译成了目的状语"以减少酸雨"。

【参考译文】
现代环境生物科技面临诸多挑战,小到通过去除电煤中的硫磺寻找可减少酸雨的微生物,大到生物降解塑料的生物生产方法。

E. g. 例如,在土木工程自己的领域里,就有下列这些分科:结构工程——研究永久性结构;水利工程——涉及水或其他流体的流动与控制系统;以及环境卫生或环保工程——研究供水系统、水净化系统与排污系统等。

【解析】原文中破折号后的信息根据逻辑关系被明示为定语从句;"涉及到水或其他流体的流动与控制系统"被解读为更清晰的英语:"which is concerned with systems involving the flow and control of water or other fluids"。

【参考译文】

For example, within the field of civil engineering itself, there are subdivisions: structural engineering, which deals with permanent structure; hydraulic engineering which is concerned with systems involving the flow and control of water or other fluids; and sanitary or environmental engineering, which involves the study of water supply, purification, and sewer systems.

2.2　英语多被动，汉语多主动

英语的被动语态可以突出动作承受着，不一定指出动作发出者（通常为人），去除了人的主观能动性，从而表意更客观。汉语往往很少直接用"被"来表示信息之间的主谓关系或动宾关系，而改用"受""遭受""受到""给""把"等被动词来表示隐性的被动意义，表面上呈现主动形态。英语的被动和汉语的主动之间的转换至关重要，否则会破坏译文的流畅感（英汉双向同理）。

E. g.　If you use the Internet regularly, your activities **are** likely **spread out** all over the Web.

你如果经常使用因特网，则你的上网活动很可能**遍及**整个网络。

E. g.　Though various Web activities can **be added** to a Live Profile, this connection isn't as productive as it could be.

虽然可以**把**各种各样上网活动**添加**到在线档案中，但是这种链接并不是十分有效。

E. g.　所有这些体系的**创建**并不是通过企业结构变化，而是通过企业经营管理和企业文化的**转变**而**实现**的。

All systems are **created** through not structural change, but process and cultural change.

E. g.　虽然在研究中可以**达到**更高的速度，但是这次测试表明商业电缆也能以更高的速度运行。

Although even higher rates **have been attained** in research, the test proves that commercial cables can also **be made to work** at higher speeds.

2.3　英语多物称，汉语多人称

与汉语相比，中、高级英语使用较多无生命事物作句子主语（即"物称"），用以客观表达事物概念。这一点对"信息型"科技文本尤为重要，因为科技文本信息更多是原理、规律的描述，极具客观性。汉语较多从人的角度出发来叙述客观事物或描述人及其行为或状态，常常使用人称词汇作句子主语（即"人称"）。

E. g.　**Opiates** have long been used by doctors to ease the pain of seriously ill or severely injured people.

长期以来，**医生们**一直采用鸦片制剂来缓解重病者或重伤者的疼痛。

E. g. **Cast iron** should not be thought of as a metal containing a single element，but rather，as one having in its composition at least six elements.
我们不应把铸铁看成是只含单一元素的金属，而应该注意到这种金属至少含六种元素。

E. g. **我们**能以各种方式给计算机发出信息或指令。
A computer can be given information or orders in various ways.

E. g. 近年来，**有人**在实验室里教黑猩猩用手势语与人交谈。
In recent years，**chimpanzees** have been taught in laboratory to use sign language for communication with people.

E. g. 在今后很长一段时间内，**我们**仍将把钢及其合金作为主要的工业材料。
Steel and its alloys will still be taken as the leading materials in industry for a long time to come.

2.4　英语为静态语言，汉语为动态语言

与汉语相比，英语中动词的使用频率低，很多动作概念用名词、形容词、副词甚至介词形式表现出来。英语写作中动词的名词化可能是这方面最典型的代表了。

E. g. Vietnamese war is a **drain** on American resource.
越南战争不断地**消耗**美国的资源。

E. g. **sentence** of life without **parole**
终身**监禁**不得**保释**

E. g. The paper gives an **overview** of recent advances in aerodynamics at DLR.
本文**概述**了德国宇航院(DLR)空气动力学的新进展。

E. g. 人的体温是靠**消耗**血液中的糖分来维持的。
Our bodies are heated by the **consumption** of sugar in the blood.

E. g. **发现**了中子之后，又发现了几种新的粒子。
Since the **discovery** of the neutron，several other kinds of new particles have been detected.

E. g. **生产**塑料不但需要科学知识，而且需要大量重型机械。
The **manufacture** of plastics requires a large quantity of heavy machines as well as a knowledge of science.

2.5　英语表态在先，叙述在后；汉语叙述在先，表态在后

E. g. **It is no longer possible** to allow the development of the full potentialities of the new metals to evolve over a period of about fifty years，as was the case with aluminum in the period between 1890 and 1949.
充分开发铝的潜力耗费了从1890年到1949年大约50年的时间，而现在要把新金属

的潜力全部开发出来**也许不再需要** 50 年那么漫长的时间了。

E. g. **It is quite obvious** from this comparison chart that quite a few minis can be used before the hardware cost will equal that of a large computer.

在硬件成本达到大型计算机成本之前，可以使用为数颇多的小型机，**这很明显**可从这个对比表格中看出。

当然，有时不一定非要这样操作，而采用顺译法。

E. g. **It is evident that** a well lubricated bearing turns more easily than a dry one.

显然，润滑好的轴承，比未作润滑的轴承更容易转动。

E. g. 然而，可以**假定**，黏性影响只局限在一层薄薄的附面层之内。

It is permissible, however, to postulate that the effect of viscosity is confined to thin boundary layer.

E. g. 眼下还**无法说**黑洞内部发生的情况。

It is impossible to say what happens inside a black hole.

2.6　英语主语显著，汉语话题显著

"主语显著"指主语和谓语是句子结构的最基本语法单位，句子一般至少有主语和谓语两个成分。英语句子有五种基本句型，这些句型由谓语中谓词用法（是否及物、是否需要双宾语等）变换而来，连同英语各种词法、语法和句法构成了英语多样化的句式。这五种句型无不彰显出英语是"主语显著"型语言。

"话题显著"指以"话题＋评论"（往往先抛出话题，然后进行评价/述评）结构构成句子表达的基本单位。汉语属于"话题显著"型的语言（赵元任，1965），"话题＋评论"这样的句式是汉语中非常常见的句式。除了主谓结构外，汉语中还有很多无主句（"话题显著"型语言的典型代表），但是这丝毫不影响汉语表达意思。

英语句子有以下 5 种基本句型：

（1）主语＋谓语。

E. g. In an engineering design, the factors of safety and economy can never be neglected.

在工程设计中，绝对不能忽视安全因素和节约因素。（汉语无主句）

E. g. Friction can be reduced by lubrication.

润滑能减少摩擦。（"话题显著"，英语主语和汉语话题相同）

（2）主语＋谓语＋宾语。

E. g. Minutes later, at an altitude of 300 miles, this tiny electronic moon begins to orbit about the earth.

几分钟后，在 300 英里的高空，这个极小的电子月亮开始环绕地球在轨道上运行了。（"话题显著"，英语主语和汉语话题相同）

E. g. Is cloning technology becoming the sword of Damocles to human beings?

克隆技术是否正日益成为人类头上的一柄达摩克利斯剑，无时无刻不在威胁着人们的安全呢？（"话题显著"，英语主语和汉语话题相同）

(3) 主语＋谓语＋间接宾语＋直接宾语。

E.g. The development of modern technology has brought human beings a lot of conveniences.

现代科技的发展给人类生活带来了很多方便。("话题显著",英语主语和汉语话题相同)

(4) 主语＋谓语＋宾语＋宾语补足语。

E.g. Larger fiber bandwidth, lower loss and more reliable optical sources would make optical fibers more competitive in this section.

由于光学纤维频带较宽,损耗较低,光源更为可靠,因此在这一领域会更具竞争力。(英语单句转变为汉语复句)

E.g. Even the dissolution in water cannot enable common salt to change its chemical properties.

即使我们将普通食盐溶于水,也不能改变其化学性质。(英语单句转变为汉语复句)

(5) 主语＋系动词＋表语。

E.g. Filtration on the industrial scale is usually underway in a centrifuge.

工业液体通常经离心机加速后进行过滤。("话题显著",英语主语和汉语话题不同)

E.g. On the continent of Europe, 55 tons was and still is a common weight for a sleeping car.

在欧洲大陆,无论是过去还是现在,一辆卧车通常重达55吨。("话题显著",英语主语和汉语话题不同)

练 习 题

一、请将下列文本译为汉语。

Foreword

Of the thousands of ways computers have been misused, simulation tops the list. The reasons for this widespread abuse are not hard to find:

1. Every simulation simulates something, but there's no particular reason it should simulate what the simulator had in mind.

2. Computer outputs are readily mistaken for gospel, especially by people who are working in the dark and seeking any sort of beacon.

3. Simulation languages have succeeded in making it easier to achieve impressive simulations, without making it easier to achieve valid simulations.

4. There are no established curricula based on extensive practical experience. Thus, everyone is an "expert" after writing one simulation, of anything, in any language, with any sort of result.

5. The promise of simulation is so great that it's easy to confuse hope with achievement.

Underlying all these reasons is lack of knowledge. Can a single book hope to remedy this kind of deficiency all by itself? Unfortunately, the answer is, and always will be, "no". The problem is too big, and books by themselves never do anything. "People" do things — sometimes with the help of good books— but books themselves are entirely passive, no matter how good they might be.

"*Simulation: Principles and Methods*" is a highly worthwhile book. Properly used, it can make a contribution to the restoration of confidence in simulation as a tool of systems thinking. That's why we've included it in our Computer Systems Series. Having put it in the Series, however, I feel obligated to say something about what I mean by "properly used".

One of the great advantages that Graybeal and Pooch have over many other authors of simulation books is their objectivity. Up until now, most of the important simulation books have been written by advocates of some particular language or approach. Such books are as much political as educational as is quite appropriate for a subject in its infancy. But we've now matured past the point where we simulate in language X or use method Y merely because we don't know anything about language A or method B. In Graybeal and Pooch, the student of simulation can obtain a balanced picture of what options are available to solve problems, rather than a treatise on "why my method is the best method of solving problems and the only method you should know."

When I refer to a "student" of simulation, I introduce another of the significant features that Graybeal and Pooch have brought to their subject—the clarity of presentation. Up until now, most people have learned about simulation through self-study. When they had a problem, they asked a friend and got a book on simulation. Because they had a problem, they were motivated to work through the book, regardless of any extraneous difficulties laid down by the author.

Today, however, there is an increasing audience of "students" who will be studying simulation in a formal class with the aid of an experienced teacher. For these students, a different type of book is needed, and "*Simulation*" addresses that need. The writing is clear, but doesn't go into more detail than a student wants when surveying a subject for the first time. There are many examples, well chosen for their pedagogic value. There are exercises with which the student can test understanding while progressing through the book. In short, "*Simulation*" is a "textbook, not simulation manual", and that's just what we demand today.

Not that "*Simulation*" shouldn't be used by anyone who's using simulation for the first time without benefit of a classroom. On the contrary, I can think of nothing better for the eager professional than to curb the impulse to plunge into some language manual, and to first read "*Simulation*" instead. The presentation is good enough that a well-motivated reader can readily glean the book's many lessons, and thus prepare for more intelligent reading which leads to the eventual manual.

  Alas, I can't require eager people to slow down and read the books they ought to read. If I could, then perhaps I could legislate some of simulations worst abuses out of existence. Lacking legislative power, I can only offer good books such as this in our series, hoping that instructors will adopt them as texts and readers will use them to evade the next generation of simulation tragedies and comedies.

  （秦荻辉. 精选科技英语阅读教程[M]. 西安：西安电子科技大学出版社, 2008.）

二、请将下列文本译为英语。

网络技术

  网络技术是从 1990 年代中期发展起来的新技术，它把互联网上分散的资源融为有机整体，实现资源的全面共享和有机协作，使人们能够透明地使用资源的整体能力并按需获取信息。资源包括高性能计算机、存储资源、数据资源、信息资源、知识资源、专家资源、大型数据库、网络、传感器等。当前的互联网只限于信息共享，网络则被认为是互联网发展的第三阶段。网络可以构造地区性的网络、企事业内部网络、局域网网络，甚至家庭网络和个人网络。网络的根本特征并不一定是它的规模，而是资源共享，消除资源孤岛。

关键技术

  网络的关键技术有网络结点、宽带网络系统、资源管理和任务调度工具、应用层的可视化工具。网络节点是网络计算资源的提供者，包括高端服务器、集群系统、MPP 系统大型存储设备、数据库等。宽带网络系统是在网络计算环境中，提供高性能通信的必要手段。资源管理和任务调度工具用来解决资源的描述、组织和管理等关键问题。任务调度工具根据当前系统的负载情况，对系统内的任务进行动态调度，提高系统的运行效率。网络计算主要是科学计算，它往往伴随着海量数据。如果把计算结果转换成直观的图形信息，就能帮助研究人员摆脱理解数据的困难。这需要开发能在网络计算中传输和读取，并提供友好用户界面的可视化工具。

面临问题

  在上篇中我们讲述了 Internet 的发展简史和它的方方面面的应用。正是由于 Internet 的丰富多彩，才会吸引越来越多的人加入其中：对用户而言，Internet 正一步步渗透到我们工作、生活的各个方面，极大地改变了长久以来形成的传统思维和生活方式；而对 Internet 而言，用户的积极参与使得这一全球通行的网络迅速膨胀而使其耐受力面临带宽的短缺、IP 地址资源匮乏等严峻考验。……

  (https://baike.so.com/doc/5682487-5895164.html)

第 3 章 科技翻译过程

今天,全世界的学者通过参加学术会议、论文发表、著作出版等方式互通有无,互相学习,共同进步,科技发展已成蓬勃之势。

科技翻译是一项重要的工作。在我国历史上,科技发展时间短,进展缓慢,直到改革开放后才得以迅猛发展,科技翻译起了重要作用。在近现代历史上,西方科技发展迅速,我国学者陆续将西方科技著作译入我国,加上我国无数科研工作者孜孜不倦、持之以恒的辛勤工作,逐渐奠定了我国科技发展的基础。

科技翻译有严格的操作流程。笔者认为,科技翻译过程可分为理解、转换、表达及修改四个阶段。其中,理解是所有翻译操作的总前提,因为如果理解错误,任何翻译操作将变得毫无意义;转换是关键,涉及从词、词组、小句、句子到段落和篇章各语言层,可以简化表达过程,是生成正确译文的重要保证;表达是译文产出阶段,需将理解并转换恰当的信息生成译文,需要译者娴熟的语体转换能力;修改同样很重要,在正确表达的基础上,修改可以产出高质量的译文。

3.1 理 解 阶 段

透彻的理解是所有后续翻译操作的总前提。

3.1.1 理解语言现象

语言是信息的载体。科技文本崇尚严谨、周密,结构有时复杂,对非此专业的译者来说,这样的文本可能读来晦涩难懂。这时,译者须借助词典、互联网等工具首先弄懂翻译对象各语言层信息。对源文本语言信息的解读能力因读者而异,双语能力是语言基本功,须下苦功夫才能掌握。

E. g. A memory is a medium or device capable of storing one or more bits of information. In binary systems, a bit is stored as one of two possible states, representing either a 1 or a 0. A flip-flop is an example of a 1-bit memory, and a magnetic tape, along with the appropriate transport mechanism and read/write circuitry, represents the other extreme of a large memory with an over-billion-bit capacity.

科技文本质朴,逻辑清晰。但是,科技范围广阔,译者不大可能了解所有领域的科技信息。语言层面的困难大多来自词汇层——词义上的确定。比如上文中,"memory""bit""flip-flop"等分别是什么意思呢?根据这里涉及的专业话题——计算机,通过查询词典得知,"memory""bit""flip-flop"分别是"存储器""比特(二进位制信息单位)""触发器"的意思。

【参考译文】

存储器是能够存储1或多比特信息的媒介或装置。在二进制系统中,1比特以两种可能状态中的一种方式存储,分别代表1或0。触发器就是1位存储器。配有适当的传送装置和读写电路的磁带是大存储器中另一个极端例子,存储能力在10亿比特以上。

E. g. The weight, measurement, marks and numbers, quality and value, being particulars furnished by the Shipper, are not checked by the Carrier on loading.

本句是简单句,呈"主语+谓语"结构,易于识别。然而,句中现在分词"being..."扮演什么功能?通过分析得知,"being particulars furnished by the Shipper"可视作后置定语,修饰其前的"The weight, measurement, marks and numbers, quality and value"整体信息,相当于定语从句"which are particulars furnished by the Shipper"。

【参考译文】

重量、尺码、标识、数量、品质和价值是托运人所提供的,承运人在装船时并未核对。

3.1.2 理解逻辑关系

英语属于形合语言,信息通过有形的词法、语法、句法、关联词、章法等连接,信息间关系处于明示状态,逻辑清晰可见。但是,汉语属于意合语言,信息间很少使用有形的连接手法明示信息之间关系。科技文本有很多专业信息,透彻地理解信息间关系是用英语清晰表达原文内容的关键。

E. g. 由于中国 AML 患者中 HLA-A*0201 比例较高,因此研究限制性的 MLAA-22 HLA-A*0201 抗原表位以诱导针对 TPH-1 细胞的特异性 CTL 具有十分重要的意义。

译者在透彻理解原文信息后,得知其核心信息是"由于……(问题),研究……以诱导……具有十分重要的意义"。本着"直译""顺译"优先原则,注意到英语表态在先(比如此处:"具有十分重要的意义"),叙事在后(比如此处:"研究……以诱导……")特点,因此,可按照本句核心信息"研究……具有十分重要的意义"搭建英语句式,一般须用 it 作形式主语。

基于以上分析,笔者将本句翻译为:For a high expression rate of HLA-A*0201 in Chinese AML patients, it has a very important significance to research the restricted antigen epitope of MLAA-22 HLA-A*0201CTL epitope for the induction of specificity CTL on TPH-1 cells.

3.1.3 理解所涉及的客观事物

《左传·襄公二十年》中有云:"言而无文,行而不远。"科技文章属于"信息型"文本,翻译时要用译入语清晰地再现原文信息,传递给译入语读者。除了源语和译入语两种语言外,译者须具备所涉及的专业知识,这样才能准确理解原文,用地道的译入语表达出原文信息。

E. g. New research notes that as quakes increased in number, so did the use of injection wells that bury wastewater from fracking and other oil and gas operations.

对于非此专业的译者而言,翻译前首先要弄懂"injection wells"是什么,"fracking"是什么意思,这样的"wells"与其后定语从句中信息是怎样的关系。诚如很多时候,科技术语须直译,"injection wells"译为"注入井",指的是在油田开发过程中,为保持或恢复油层压力,在油田边缘或内部钻凿的往油层中注水或注气的井。"fracking"是专业术语,意思是"水力压裂"。根据句中信息之间的逻辑关系,"injection wells that bury wastewater..."这种井用于注入油田上作业时产生的废水。

所以笔者将本句翻译为:新的研究发现,随着地震次数增加,开采石油和天然气时用于排放因水力压裂及其他作业而产生废水的注水井数目也在增加。

3.2 转换阶段

转换,即英语的"transfer"一词,是译者在很强的语体差异意识下,按照译入语的行文习惯做出的文本调整。转换往往按照自下而上的顺序进行,即按照语音—词汇—小句—句子—段落—篇章这个顺序进行,信息层级越来越高,语言单位越来越大。

具体而言,各层级信息转换的做法如下:

(1)在语音层,译者一般要音译,措辞符合音译规范,如有现成译法,须与权威部门保持一致;

(2)在词汇层,译者要做到措辞准确,术语翻译与国际标准、国家标准、行业标准保持一致;

(3)小句,即一个无动词结构并包含主谓关系的成分,要按其功能,结合上下文做出恰当的翻译转换(译作单句、从句还是其他某种形式);

(4)在句子层,译者要注意查看原句与译入句结构差异,对有差异情形要做出适当、合理的调整,最终搭建起既符合译入语表达习惯,又给予句中其他信息附着余地的句式;

(5)在段落层,译者要注意段落信息的完整性、段落主题的凸显性、句子之间的衔接性以及段落中信息的推进性,使得译文段落如原文段落一样信息饱满,浑然一体,读来通畅;

(6)在篇章层,译者要按照译入语文章模式架构译文结构,同时也要兼顾(1)~(5)层信息的恰当转换,使得译文如同原作,达到钱钟书先生所说的"化境"。

"化境"说法出自于钱先生于1981年发表的《林纾的翻译》一文中。在这篇文章中,钱先生指出,"文学翻译的最高标准是'化'。把作品从一国文字转变成另一国文字,既能不因为语文习惯的差异而露出生硬牵强的痕迹,又能完全保存原有的风味,那就算得入于'化境'。"这样的译本,按钱先生的看法,应该"忠实得以至于读起来不像译本,因为作品在原文里决不会读起来像经过翻译似的。"

3.3 表达阶段

这个阶段生成第一稿译文。因为转换阶段原文各信息层已转换到位,所以理想中的表达阶段困难一般不大。译者在表达时,要动态监控语言各层面源语与译入语的信息对等情况,对于"对等"效果不佳的情形要立即做出调整。

笔者坚信,笔译是严肃的目的语写作,写作几乎所有的要求适用于笔译。因此,一位好译者应该也是写作高手。表达时,译者不能仅仅满足于译文"忠实"和"通顺",要力求在保证字正句顺的基础上,用自己能驾驭对的更高层面的语言重现原文信息。这样,读者阅读译文后不但获取了原文信息,而且获得了译文文字美的感受,在一定程度上提高了自己的写作水平,甚至在思想上与原作者产生共鸣。

3.4 校 对 阶 段

为保证译文质量,校对是必不可少的阶段。在这个阶段,译者对原文内容(甚至风格)在译文中的再现情况进一步核实,对译文语言推敲、润色,以求在内容上与原文一致,在形式上进一步达到完美。这个阶段对科技翻译尤为重要,因为自然科学文本中有公式、数据等关键信息,切不可译错。

核对内容可参照转换阶段各信息层的转换内容,逐层、多次检查,力求没有错误。

练 习 题

一、请将下列文本译为汉语。

Shenzhou Ⅵ Spacecraft

Sz–6 mission marks start of manned space experiments

China's planned launch of Shenzhou Ⅵ vessel, its second manned space mission, signals that the country begins to carry out aerospace experiments with real-human participation, said a senior space engineer here Sunday. Wang Yongzhi, chief general designer of China's Manned Spaceflight Program, said that the Shenzhou Ⅵ vessel will enable astronauts to do scientific experiments in space, which offers a unique vacual, highly radiant and low gravity environment to carry out scientific studies.

China became the third nation to succeed in manned space flight when it launched the Shenzhou Ⅴ in October 2003, carrying sole astronaut Yang Liwei around the earth 14 times, who did not leave his seat in the return module during his one-day flight. Wang said that following Shenzhou Ⅵ's flight, China has greater plans such as astronauts' spacewalk, the docking of capsule with space module, launch of space lab and setting up space flight, the most complicated and difficult aerospace project, demonstrates a nation's scientific research and economic strength. He said, "It's a major means to expand human living space and tap and use space resources. China will never be a superpower, but as the world's biggest developing country with 1.3 billion people, it should have a place in aerospace development and make due contributions."

Successful Launching

China sent its second manned spacecraft, Shenzhou-6 carrying astronauts Fei Junlong and Nie Haisheng, into space on Wednesday morning for a spaceflight that lasted several days. At 9:00 a.m., Shenzhou-6 blast off from the Jiuquan Satellite Launch Center in northwest China atop a Long March Ⅱ F carrier rocket. The launch was declared a success 39 minutes later. Chinese leaders, including President Hu Jintao and Premier Wen Jiabao, watched the launch. It is estimated hundreds of millions of Chinese people watched live TV broadcast, radio broadcast or Internet coverage of the launch, further proof of China's rapidly growing scientific and technological progress. With the launch of Shenzhou-5, China became the third country in the world to carry out a manned spaceflight after the Soviet Union and the United States two years ago. Unlike the maiden manned spaceflight in October 2003, which carried astronaut Yang Liwei into space, this mission tested China's capability in conducting spaceflight that carried more than one person and lasted more than one day. It also involved "scientific experiments in space with true human participation" that enhanced the level of the experiments. The spacecraft had been in good condition since the launch. It changed the previous elliptical orbit to a circular at 3:50 p.m. The astronauts successively entered the orbital capsule from the return capsule to conduct experiments about two hours later...

(秦荻辉. 精选科技英语阅读教程[M]. 西安:西安电子科技大学出版社,2008.)

二、请将下列文本译为英语。

随着我国城镇化的快速发展,集约高效、节能环保的公共交通已成为构建资源节约型及环境友好型社会的战略选择。从2011年起,我国大力推动创建"公交都市"的重要举措,鼓励发展快速公交(BRT)等地面大运量公交,强调"提升公共交通设施和设备水平"。BRT以公交车为基础,采用隔离专用道等高质量基础设施及快速频繁的运营,提供快速、舒适、低成本的公交服务。有别于常规公交站的简易站亭,BRT车站的组成复杂,规模大,既是乘客体验公交服务的重要场所,也是城市道路系统相对独立的建筑设施。车站的形式和布局影响系统运行的效率,站内环境的舒适性影响系统的吸引力,因此BRT车站建筑设计是系统质量的决定性因素之一。据统计,全球已有超过160个城市实施了BRT,其中我国有近30个城市。然而,现有的文献比较缺乏对BRT车站建筑的研究,特别是车站的热舒适性与节能方面,本文试图通过分析现有案例,探讨BRT车站建筑的气候适应性设计的主要策略。

1. BRT车站建筑的特点

第一,BRT车站的组成较复杂,一般包含设备房、售检票亭、检票闸机及候车登乘区。为减少登乘时间,BRT采用地铁系统的登乘前售检票方式,因此须设隔离的候车区域。

第二,车站的形式、位置及规模主要由系统运营模式、客流需求、线路数量及道路空间决定。车站设计应便于系统的运营和管理。

第三,车站规模大。车站长度通常超过30m,而且登乘区会根据子站数量的增加和停靠超长公交车的情况而增长。车站宽度取决于结构宽度及客流量,一般超过3m。站台高度与车辆地板高度一致,便于水平登乘。

第四,车站内部环境受周边道路交通影响大。为了提高车辆运行效率,BRT专用走廊多在路中,沥青路面温度高,汽车尾气污染严重,且周边缺乏绿树遮阳,因此车站所处环境的热舒适性较差。车站必须设有气候保护设施,提升候车环境的舒适性。

第五,相比其他大运量公交,BRT的优势在于成本低、实施时间短。因此,BRT车站追求低成本、易实施和标准化。建筑材料多选用钢、混凝土、玻璃等常用材质,安全耐用,容易实施装配。但这些材料容易大量吸收热辐射,增加站内温度。另外,为降低成本和能耗,车站较少采用空调、暖气等调节气候的机械设备,因此车站设计更注重通过建筑形式和构件的选择"被动地"调节站内热环境。……

(黎淑翎,周剑云.快速公交车站建筑气候适应的主要策略初探[J].华中建筑,2018(11):124-128.)

第4章 科技翻译常用的方法

科技翻译常用的方法有音译、照译、增译、省译、词类转换、正反译、结构调整等。这些方法与直译、意译、顺译、倒译、拆译、合译等方法既有重叠之处，又有不同之处。

翻译方法是翻译操作的技巧性路径，对翻译工作者（尤其是翻译初学者）起着事半功倍的作用。在翻译过程中，译者须心怀这些常用的翻译方法，利用娴熟的双语应用能力，做到两种语言间恰当的转换即可开展翻译工作。

4.1 词和一般句子的译法

4.1.1 音译

音译，顾名思义，根据原词发音做出翻译操作。

这一点译界有严格的规定。翻译所要用到的字眼须参照专有名词词典、官方文件等材料与权威机构和官方说法保持一致。近现代，西方发达国家在科技很多领域超越了中国，很多科学规律的发现、科学技术的发明及其命名都是由西方国家执行的，然而这些名称在汉语中没有对应词，所以一般情况下通过音译出现在汉语中，一定程度上丰富了汉语语言，为汉语语言的发展做出了贡献。

E.g. calorie 卡路里；nylon 尼龙；sonar 声呐；quark 夸克；celluloid 赛璐珞；radar 雷达；gene 基因；hertz 赫兹；nanometer 纳米；Celsius 摄氏；sauna 桑拿浴；system 系统；tank 坦克；tire 轮胎；mosaic 马赛克；lumen 流明（描述光通量的物理单位）；Pentium 奔腾；等等。

4.1.2 照译

照译，即按照原文字面含义进行翻译，与直译类似。

科技英语中大量的术语是照译到汉语中的。

E.g. acid fume 酸雾；acoustic analysis 声学分析；access method 存取方法；account number 账号；air brake 气闸；alarm lamp 报警灯；back reaction 逆反应；bank deposit 银行存款；bar code 条形码；biological concentration 生物浓缩；binary logic 二进制逻辑；chained record 链式记录；centrifugal separation 离心分离；cross walk 人行横道；cruising endurance 续航力；crystal glass 晶体玻璃；data flow chart 数据流图；dust cloud 尘团；hybrid reasoning 混合推理；humidity sensor 湿度传感器；human-caused error 人为误差；man computer communication 人机通信；manned orbital space station 载人轨道空间站；manual entry 人工录入，手控输入；perfect information feedback 全信息反馈；rocket motor assembly 火箭动力装置；robotic artificial intelligence 机器人人工智能；Deoxyribonuclear acid＝DNA＝de＋oxy＋ribo＋nuclear acid 脱氧核糖核酸；stem cell 干细胞；optical drive 光驱；satellite antimissile observation system 卫星反导弹观察系统；anti-armored-vehicle-missile 反装甲车导弹；等等。

E. g. Physics studies force，motion，heat，light，sound，electricity，magnetism，radiation and atomic structure.
物理学研究力、运动、热、光、声、电、磁、辐射和原子结构。

E. g. Different forms of energy all can be used to do work.
各种不同形式的能都可用来做功。

4.1.3 增译

英语讲究简洁、明快，要尽量避免重复。相对而言，汉语不讨厌重复，因为词语反复使用有时不但有助于语义显现，而且有节奏乐感，给人以美感。而且，汉语倾向于明示话题所涉及的范围或领域，比如"状况""现象""结果""作用"等。因此，英译汉时，译者须增加这些表示范围和领域的词才能清晰表意。

以下例子中粗体字都是增加的信息。

E. g. abortion 堕胎**做法**；distribution 分布**情况**；upheaval 动乱**局面**；juvenile delinquency 少年违法**犯罪**；prosperity 繁荣**景象**；mortality 死亡**数**，死亡**率**；等等。

E. g. In rapid oxidation a flame is produced.
在快速氧化**过程**中会产生火焰。（在抽象名词"氧化"后增补"过程"使所指更明确）

E. g. Air pressure decreases with altitude.
气压随海拔**的升高**而降低。

E. g. This set of records are cheap and fine.
这套唱片**物美价廉**。

E. g. Clearly, cool temperature slows down the action of bacteria.
显然，低温**使**细菌的活动减慢。（对动词增补"使"字，把原文的动宾结构转译成汉语里常用的递进结构）

E. g. Note that the words "velocity" and "speed" require explanation.
请注意，"速度"和"速率"这两个词**需要解释**。

4.1.4 省译

省译，即翻译时有省略情形。

E. g. Radar involves transmitting electromagnetic waves of very short wave length.
雷达发射的是波长非常短的电磁波。（动词"involves"略去未译，对主语等成分作了必要的转换）

E. g. Like charges repel each other while opposite charges attract.

同性电荷相斥,异性电荷相吸。(省译"while")

E.g. A gas distributes itself uniformly throughout a container.
气体均匀地分布在整个容器中。(省译"itself")

E.g. 质子带正电,电子带负电,而中子**既不带正电**,**也不带负电**。
A proton has a positive charge and an electron a negative charge, but a neutron has neither.

4.1.5 词类转换

词类转换,即在翻译过程中词性发生了变化,如动词译为名词,形容词译为动词,副词译为形容词。由于英语和汉语差异大,词类转换也许是使用最频繁的一种翻译方法。

E.g. Development of the internal combustion engine took place largely during the 19th century.
内燃机很大程度上是在19世纪**发展**起来的。(名词转换为动词)

E.g. China's first atomic **blast** in October 1964 was a great **shock** to the world.
1964年10月,中国**爆炸**了第一颗原子弹,全球为之**震惊**。(名词转换为动词)

E.g. The **discovery** of America is generally referred to Columbus.
一般认为美洲是哥伦布**发现**的。(名词转换为动词)

英译汉时,将表示动作概念的名词译为汉语动词,究其根本原因是英语和汉语差异导致的:英语是静态语言,汉语是动态语言(参第2章英汉差异部分)。

E.g. They have been **victimized** by the experiments.
它们成了这些试验的**牺牲品**。(动词转换为名词)

E.g. Coating thicknesses **range** from one tenth mm to 2 mm.
涂层的厚度**范围**是0.1～2 mm。(动词转换为名词)

E.g. Text analysis can extract a set of keywords that **characterize** the document.
文本分析可以抽取出一组代表文档**特征**的关键词。(动词转换为名词)

E.g. She's no less **capable** than the others.
她的**能力**并不比别人弱。(形容词转换为名词)

E.g. Below 4℃, water is in **continuous** expansion instead of **continuous** contraction.
水在4摄氏度以下就**不断地**膨胀,而不是**不断地**收缩。(形容词转换为副词)

E.g. The temperature in the furnace is not always **above** 1,000℃.
炉膛内的温度并不总是**处在**1 000℃以上。(介词转换为动词)

4.1.6 正说反译/反说正译

"正说反译",即直观地看,原词/文以肯定形式表意却翻译成了否定形式;"反说正译"恰恰相反,即原词/文以否定形式表意却翻译成了肯定形式。

E.g. Admittedly, much of what we're prescribing dovetails with the recommendations of the Iraq Study Group.
无可否认,我们所开的处方有很多地方与伊拉克研究小组的建议吻合。(正说反译)

E.g. Whenever possible, **avoid** working on a running engine.

切勿在发动机处于运行状态时进行维修。（正说反译）

E. g. All materials conduct an electric current **to some extent**, even though some (insulators) have very high resistance.

即使某些材料(绝缘子)的电阻很大,但是所有的材料都能导电,只是**程度不同**而已。（正说反译）

E. g. In one major form, juvenile-onset diabetes, the pancreas supplies **little**, if any, insulin.

有一大类是青少年糖尿病,胰腺**几乎不**分泌胰岛素。（正说反译）

E. g. It was **unfortunate** that opening this case, we found that owing to negligent packing several attachments were damaged to such an extent that they can not be used directly.

非常遗憾,开箱后发现,因包装疏忽,致使几个零件损坏,已无法直接使用。（反说正译）

E. g. There is **hardly** any scientific field that does not deal with cause and effect.

几乎没有一个科学领域不和因果有关。（反说正译）

E. g. **No** smoking!

禁止吸烟！（反说正译）

4.1.7 结构调整

英语和汉语句子结构存在较大差异。英语句子往往呈现"主语＋最重要部分信息＋废话（相对而言不重要的信息,如状语）"结构,而汉语句子(除无主句外)常常以"主语＋废话（相对而言不重要的信息,如状语）＋最重要部分信息"结构呈现信息。因此,为了清晰、自然、地道表意,译者时不时要做出结构调整。

E. g. These stations may report weather at intervals of an hour or even less to warn of approaching weather dangers.

这些气象站可以每隔1小时或不到1小时报告1次天气,以警告人们危险天气即将来临。

E. g. Steel is widely used in industry because it possesses a great number of useful properties.

钢因为具有许多有用的性能而在工业上得到广泛应用。

E. g. These voice messages can be accessed later by the person to whom they are addressed.

信息接收人可以晚些时候收到这些语音信息。

E. g. Last week, scientists completed the first stage of this remarkable transfer, using a helicopter to lift a 23-ton block of ice and mammoth to a new site where defrosting can be started.

上周,科学家使用直升机抬起重达23吨的猛犸冰块,将其运至一个新的地方解冻,完成了举世瞩目的第一阶段工作——转移猛犸躯体。

E. g. 物理学是研究物质世界最基本的结构、最普遍的相互作用、最一般的运动规律及所使用的实验手段和思维方法的自然科学。

Physics is a discipline of natural science that uses experimental methods and thinking modes to study matter, its most basic structures, its most common forms of interactions and its most general laws of motion.

E.g. 经过大量严格的实验验证的物理学规律被称为物理定律。

Physical rules which are validated by a number of strictly controlled experiments are termed "laws of physics".

4.2 科技语篇中长句的译法

由于科技文章要求对客观事物的描述要准确、完整，句子层次分明，体现思维严谨、逻辑缜密的文风，所以在科技英语中有许多由从句、非谓语动词以及各种修饰语等组成的长句。长句是翻译难点之一，其翻译质量对整篇文本翻译质量有很大的决定作用。

翻译长句时，首先要分析原文的句法结构，弄懂整句的中心意思，理清信息层次及信息之间的逻辑关系，然后根据信息之间的相互逻辑关系（因果、条件、让步、时空顺序等），按照目的语的特点和表达方式，正确地译出原文的意思。

长句的翻译分5步进行：

(1) 划清句子成分。根据英语词汇在英语句子中的地位和作用，英语句子的成分可分为主语、谓语、宾语、宾语补足语、表语、定语、状语、同位语及独立成分等。现代汉语的句子成分一般有八种，即主语、谓语、宾语、动语、定语、状语、补语和中心语。

(2) 弄懂整句的中心意思。

(3) 理清信息层次及信息之间的逻辑关系。

(4) 逐一翻译每句各部分信息。

(5) 对译文进行调整、组合，然后检查、修改、润色，完成译文。

4.2.1 顺译法

顺译法，即按照原文信息罗列顺序生成译文。当英语和汉语表达顺序大体相同时适用此方法。

E.g. Cases of Legionnaire's disease are becoming fewer with newer system designs and modifications to older systems, but many older buildings, particularly in developing countries, require constant monitoring.

莱金奈尔病病例减少了，因为（空调）系统设计更新了，而且对旧系统也进行了改进，但是很多老旧楼房（尤其是在发展中国家）还需要持续监控。

E.g. Plastics is made from water which is a natural resource inexhaustible and available everywhere, coal which can be mined through automatic and mechanical processes at less cost and lime which can be obtained from the calcinations of limestone widely present in nature.

塑料是由水、煤和石灰制成的。水是取之不尽的、到处可以获得的天然资源；煤是用自动化和机械化的方法开采的，成本较低；石灰是由煅烧自然界中广泛存在的石灰石得来的。

E.g. 随着用来制成这些芯片的元件变越来越小,宽度从几百纳米变成几十纳米,组装它们变得更加困难。

As the components used to make these chips become smaller, with widths measured in dozens rather than hundreds of nanometers, they become increasingly more difficult to assemble.

E.g. 了解一些历史会有所帮助,但是掌握生物学知识现在更有用的——例如阿米巴虫是如何不断一分为二、分裂繁殖的。

It helps to know some history, but a grasp of biology now is even more useful — like how an amoeba reproduces by constantly splitting itself in half.

4.2.2 倒译法

这是由英语和汉语句式构建的差异导致的。英语句式多将主要信息向句首靠近罗列,呈"主语+谓语+废话(其他相对不太重要的信息,比如状语)"结构,而汉语呈"主语+废话(其他相对不太重要的信息,比如状语)+谓语"或"废话(其他相对不太重要的信息,比如状语)+主语+谓语"结构。

E.g. This was found to have affected people in buildings with air-conditioning systems in which warm air pumped out of the system's cooling towers was somehow sucked back into the air intake, in most cases due to poor design.

研究人员发现,空调系统压缩机排出的热气不知怎的(大多由于设计不当)吸入进气口,从而影响建筑物中人们的健康。

E.g. It is a mechanical means that a float switch is usually used to turn off the burner or shut off fuel to the boiler to prevent it from running once the water goes below a certain point.

当水位低于某个点位时,使用浮控开关关掉燃烧器或断开锅炉的燃料供应,防止锅炉继续运行,这是一种机械手段。

E.g. Nearly any material can be used for lighting fixtures, so long as it can tolerate the excess heat and is in keeping with safety codes.

只要能够耐受过热并符合安全标准,几乎所有材料都可以用于照明灯具。

E.g. Human beings have distinguished themselves from other animals, and in doing so ensured their survival, by the ability to observe and understand their environment and then either to adapt to that environment or to control and adapt it to their own needs.

人类具有能力观察和了解周围环境,然后要么适应环境,要么控制环境并根据自身的需要改造环境。就这样,人类把自己和其他动物区别开来,一代代生存下来。

E.g. 由于中国 AML 患者中 HLA－A＊0201 比例较高,因此研究限制性的 MLAA－22 HLA－A＊0201 抗原表位以诱导针对 TPH－1 细胞的特异性 CTL 具有十分重要的意义。

For a high expression rate of HLA-A＊0201 in Chinese AML patients, it has a very important significance to research the restricted antigen epitope of MLAA-22

HLA-A*0201CTL epitope for the induction of specificity CTL on TPH-1 cells.

4.2.3 拆译法

拆译法,顾名思义,就是将原文中一部分信息从主句中拆出来翻译的方法,也称分译法。这主要是因为原文信息"臃肿"(如从句套从句情形),句子长,结构复杂。翻译时将从句或短语拆开来翻译,以保证译文读来流畅,逻辑清晰,层次分明。这种方法多见于英译汉情形。

E.g. The ancients tried **unsuccessfully** to explain how a rainbow is formed.
　　古人曾试图解释虹是怎样形成的,**但没有成功**。(拆译副词)

E.g. His recommendation **that the Air Force should investigate the UFO sighting** was approved by the commission and referred to the appropriate committee.
　　他建议**空军调查不明飞行物的情况**,这一建议得到了调查委员会的批准,并提交专门委员实施。(拆译同位语)

E.g. 对Internet而言,用户的积极参与使得这一全球通行的网络迅速膨胀而使其耐受力面临带宽的短缺、IP地址资源匮乏等严峻考验。
　　As for the Internet itself, active user participation is causing it to expand so rapidly on a global scale that its endurance capacity faces severe challenges, such as a shortage of bandwidth and IP address resources. (拆译定语)

有时,拆译法也指将原文一个长句拆为译文2~3句的翻译方法,与合译法相反。

E.g. The reader might speak into a telephone where the information is transduced from patterns of compressed air molecules traveling at the speed of sound into electronic pulses traveling down a copper wire closer to the speed of light.
　　读者可以通过电话交谈。这时,信息从以声速传输的压缩空气分子模式转换为沿着铜导线以近似光速传播的电子脉冲模式。(原文1句拆为译文2句)

E.g. 现代计算机系统小到微型计算机和个人计算机,大到巨型计算机及其网络,形态、特性多种多样,已广泛用于科学计算、事务处理和过程控制,日益深入社会各个领域,对社会的进步产生深刻影响。
　　Modern computer systems, ranging from microcomputers and personal computers to giant computers and their networks — with their various forms and characteristics — have been widely used in scientific computing, transaction processing and process control. **With each passing day, modern computer systems are penetrating ever more deeply into all areas of society and are thus having a profound impact on society's progress.** (原文1句拆为译文2句)

4.2.4 合译法

合译法,指的是将原文2~3句(或更多句)话译为译文1句话的翻译方法。这种翻译方法多见于汉译英情形。

E.g. 就像产品经理常说的那样,"我们要保证有足够的利益进行产品的售后服务和研发环节"。从全球发展的角度来看,确实如此。
　　As the product manager often says, "We want to ensure that there is enough profit

reserved to improve after-sale services of our products and promote research and development work," **which is true from the perspective of global development.**（原文 2 句合译为译文 1 句）

E.g. 前者是借助电、磁、光、机械等原理构成的各种物理部件的有机组合，是系统赖以工作的实体。后者是各种程序和文件，用于指挥全系统按指定的要求进行工作。

The former, which is the entity on which the system works, is an organic combination of various physical components based on the principles of electricity, magnetism, light and machinery; the latter is a variety of programs and files used to direct the whole system to work according to specified requirements.（原文 2 句译为译文并列形式，用分号连接，变为 1 句）

4.2.5 插译法

插译法就是用添加破折号或括号的方法将部分信息合理插入到译文句子中的翻译方法。

E.g. Changing attitudes among all voters, **and especially Democratic** voters, made support for same-sex marriage an article of faith for anyone seeking to lead the party.

所有选民（**特别是民主党选民**）态度的变化使所有候选人相信，若想领导民主党，就必须支持同性婚姻。

E.g. The Republicans shifted their views from 2007 through 2011, the early years of the Obama presidency.

共和党人的态度曾在 2007 年至 2011 年期间（**即在奥巴马就任总统的最初几年**）发生转变。

E.g. 北京时间 2016 年 9 月 8 日凌晨 1 点，2016 苹果秋季新品发布会于在美国旧金山的比尔·格雷厄姆市政礼堂举行。

At 1:00 a.m. on Sep. 8, 2016 (**Beijing time**), 2016 Apple's Fall Launch Event was held in the Bill Graham Civic Auditorium in San Francisco.

4.2.6 重组法

重组法，即将信息进行重新组合的方法。

有时，原文信息成分繁多，层次复杂，无论采用顺译还是倒译，译文逻辑关系都显得比较凌乱，读来不够通顺。这时，译者就需要在充分理解原文的基础上打乱原文信息，按照译入语的表达习惯重新安排信息顺序。

E.g. Should people support genetically modified food? The debate has been ongoing across the world in recent years, as global climate change has led to a series of disastrous weather conditions that are reducing crop yields, and turning food supply into a global issue.

我们应该支持食用转基因食品吗？近年来，在世界各地，全球气候变化导致了一系列灾难性天气状况，农作物产量减少，食品供应变成了一个全球性问题，这一争论一直在持续中。

E. g. After leaving Clarksville, Tennessee, on September 6, 1916, Saunders launched the self-service revolution in the USA by opening the first self-service Piggly Wiggly store, at 79 Jefferson Street in Memphis, Tennessee.

1916年9月6日,桑德斯离开田纳西州克拉克斯维尔市,在孟菲斯市杰佛逊大街79号开设了第一家皮格利·威格利(Piggly Wiggly)自助服务店,开启了美国自助服务革命运动。

E. g. 人们普遍认识到供水和卫生设施可以改善家庭成员的身体健康状况,使妇女可以支配更多的时间。然而事实上基础设施给妇女带来的好处往往远不止这些。

The beneficial impacts of infrastructure on women can be profound, often extending beyond the commonly cited impacts of water and sanitation infrastructure on household health or women's time allocation.

E. g. 科学态度,即研究和运用物理学、生物学、化学、地质学、工程学、医学以及其他科学的男女科学工作者的态度,其本质是什么?

What is the nature of the scientific attitude, the attitude of the man or woman who studies and applies physics, biology, chemistry, geology, engineering, medicine or any other sciences?

练 习 题

一、请将下列文本译为汉语。

Following asemiotically based concept of culture, this article presents an overview of the metaphoric conceptualizations of the colors black and white based on rich cross-linguistic empirical data from the BNC and Corpus de Referencia del Español Actual (CREA). There is a mixture of literal and figurative meanings in these two comparable corpora but the focus of attention is just on figurative meanings, relating multiword units such as collocations, idioms or proverbs to their different cultural contexts.

The general research question is: how and to what extent does culture actually show up in metaphors and metonymies related to black and white and their Spanish counterparts negro and blanco? To answer this question, Piraiinen's taxonomy (Piirainen E, Phrasemes from a cultural semiotic perspective. In Burger H et al(eds), pp 209 - 219, 2007) will be used, in order to analyze different kinds of cultural phenomena from a qualitative point of view. The results show that cultural metaphors appear to require an understanding of the input domains and their properties or connections with the output domains.

The comparative outline of phrasemes containing black/negro and white/blanco clearly indicates the cultural foundation of phraseology (Wierzbicka, Semantics-primes and universals. Oxford University Press, Oxford, 1996). The uses of black/negro as "bad, unhappy" and white/blanco as "good, innocent" represent cultural facts and if taken as physical entities (color terms), they symbolize these properties. English and Spanish "black" and "white" collocations, idioms and proverbs are powerful symbols in culture. The amount of knowledge that language users have on the relationship between the symbols BLACK and WHITE in language and culture allows that the "right" reading can be activated in different contexts.

(PLAZA S M. Black and White Metaphors and Metonymies in English and Spanish: A Cross-Cultural and Corpus Comparison [C]//TRILLO J R. Yearbook of Corpus Linguistics and Pragmatics 2015: Current Approaches to Discourse and Translation Studies. Switzerland:Springer International Publishing,2015:39-63.)

二、请将下列文本译为英语。

道 歉 函

××市兴达贸易有限公司:

  贵公司20××年×月×日函收悉。函中所诉20××年1月7日《购买电脑桌合同》中,所收的35套黄花牌电脑桌部分出现接口破裂一事,深表歉意,此事已引起我方高度重视,现已就此事进行调查。

  经有关部门查实:我厂生产的××××型黄花牌电脑桌,出厂时,经质检部门检验全部为优质产品。函中所提的部分电脑桌出现接口破裂,是由于我方工人在出仓时搬运不慎造成的。对贵公司的损失,我公司再次深表歉意,并请贵公司尽快提供电脑桌受损的详细数字及破损程度,以及公证人证明和检验证明书,我公司将以最快的速度按实际损失给予无条件赔偿。

  对此,我们将引以为戒,查找工作中存在的问题和不足,制定改正措施杜绝此类事情再次发生。

  希望能够得到贵公司谅解,继续保持良好的贸易往来关系。

候复

×××市光明家具有限公司

(https://wenku.baidu.com/view/4b409841852458fb770b56a6.html###)

第5章　科技英语专有名词及术语翻译

5.1　科技英语专有名词的译法

专有名词指的是特定的或独一无二的人或物,大致包括人名、地名、国家名、商标名、江河湖海名、书名、标题、歌曲名、机构名、日期、节日名或者某一事物所专有的名词。

在科技文本中有很多专有名词。科技英语词汇大多借用常用词汇,在某个特定领域词义明显单一,词源多为希腊语和拉丁语,缩略语使用广泛,前后缀出现频率高。

专有名词的翻译要符合国际标准、国家标准和行业标准。在这方面,前人已经为我们做出了很大的贡献,编纂了各类专有名词翻译词典。翻译碰到专有名词时,不要随心所欲,而要仔细搜索、查询并采用官方已钦定的译法,与权威部门的译法保持一致。

在译名问题上,我国新华社译名室始终遵循"名从主人、约定俗成"的原则,包括专有名词在内,至今已钦定了不计其数的译名。准确地说,专有名词的翻译与译者的检索能力、语际转换意识以及背景综合知识密切相关,与一般意义上信息语际转换能力没有必然的联系。

5.1.1　人名和地名的翻译

1. 英译汉

外国人名和地名内涵意义相对较少,只是和所指称的对象形成一一对应的一个符号。译入汉语时,语义不是翻译转换的重点,所以通常采用音译法。

E. g. Barack Obama 巴拉克·奥巴马;George Bush 乔治·布什;George Washington 乔治·华盛顿;Benjamin Franklin 本杰明·富兰克林;Thomas Jefferson 托马斯·杰斐逊;New York 纽约;Pennsylvania 宾夕法尼亚;California 加利福尼亚;等等。

翻译外国人名和地名时要注意以下两点:

(1)确定该译名是否已经在汉语中"约定俗成",如果官方机构已译出,应采用官方译法。

比如:美国时任总统的姓"Trump",若按英语发音,汉译为"川普"应该更准确点。但是,新华社译名室很早以前就将"Trump"译为"特朗普"了,为避免混乱,所以沿用了旧译。根据《英汉音译表》,Newton 若被译为"纽顿"似乎比"牛顿"更为规范;按照《法汉音译表》,根据发音,用"德卡尔特"译 Descartes 会优于"笛卡尔",但"牛顿"和"笛卡尔"在中国早已广为人知,因此必须继续沿用下去(康志洪,2012)。

(2)翻译前应确定名称持有者的国籍或民族身份,尽量遵从其所属民族语言的发音规律,须在《英汉音译表》《法汉音译表》《德汉音译表》等不同语体转换音译表中查询对应的译法。

比如:俄罗斯罗曼诺夫王朝第一位沙皇的英文译名是"Michael I",根据俄语译作"米哈伊尔一世",而不能译作"迈克尔一世";法国前总统 Charles Dt Gaule,根据《法汉音译表》,译作"夏尔·戴高乐",因为"Charles"法译汉为"夏尔",英译汉为"查尔斯",比如 Charles Darwin 汉译为"查尔斯·达尔文"。

再比如:英国城市 Cambridge 汉译时采用"音译+直译"的方法,译为汉语"剑桥";作为美国城市,却采用音译,译作"坎布里奇(美国马萨诸塞州城市)";Scarborough 作为多伦多一个区的名称应译作"士嘉堡",而作为英格兰一自治市的名称应译作"斯卡伯勒"。

2. 汉译英

我们中国人名译为英语时,一般不用威妥玛式拼音,而用汉语拼音。具体做法是:姓和名分开,第一个字母大写,名字拼音写在一起,先姓后名多见,或先名后姓。

E. g. 闻一多 Wen Yiduo;张立文 Zhang Liwen;李开复 Li Kaifu

如果已有约定俗成或广为流传的固定写法,应继续沿用。

E. g. 孔子 Confucius;孟子 Mencius;孙中山 Sun Yat-sen;蒋介石 Chiang Kai-shek;等等。

地名的翻译。比如:"西安市友谊西路 127 号"则按英语小地址在前、大地址居后的习惯用英语书写为 127, the Youyi Rd, Xi'an,其中的"友谊"按现行的规范用汉语拼音即可,因为对外国人而言,这只是一条路的名字而已,不必将此处的"友谊"译作 Friendship。

5.1.2 机构名的翻译

1. 英语和汉语之间机构名称的相互转换

英语和汉语之间机构名称的相互转换常常需要直译,有时是音译。

E. g. 中华人民共和国教育部 Ministry of Education of the People's Republic of China;中国银行 Bank of China;科学院 academy of sciences;中华人民共和国商务部 Ministry of Commerce of the People's Republic of China;华为 Huawei;海尔 Haier;United States Congress 美国国会;British Parliament 英国议会;The White House Office 白宫办公厅;Coast Guard 海岸警卫队;等等。

有些机构名是通过跨语言重新命名方式进入到目的语的。

E. g. 联想 Lenovo;宝马汽车 BMW(Bavarian Motor Works);HSBC(Hong Kong and Shanghai Banking Corporation)汇丰银行;等等。

在这里,原文和译文表面上看没有关系。

2. 涉及第三语言的机构名称

在英语中,除汉语以外的第三语言机构名称往往是以音译形式或原文的文字形态直接借用的(康志洪,2012)。

E. g. FIFA 国际足球联合会(简称"国际足联",源自法语名称来 Fédération Internationale de Football Association);Al Qaeda 基地组织(恐怖主义组织,音译,源自阿拉伯语);Aum Shinrikyo 奥姆真理教(音译,源自日语);等等。

总之,汉译英机构名时,不可一味地根据意思将第三语言机构名译入英语,这样就可能译出一个英语中根本没有实际指称对象的名称。

5.1.3 媒体、刊物名与文献、著作名的翻译

1. 媒体、刊物名的翻译

一般情况下,媒体和刊物名称大都采用直译、音译或原文借用的方式进入译文中。

（1）直译。

E. g. （以下例子来自"百度文库"）：

英国主要报纸的翻译：

Times《泰晤士报》；*The Guardian*《卫报》；*The Financial Times*《金融时报》；*The Daily Telegraph*《每日电讯报》；*The Observer*《观察家报》；*The Daily Express*《每日快报》；*The Daily Mail*《每日邮报》；*The Mirror*《镜报》；等等。

美国主要报纸的翻译：

The Los Angeles Times《洛杉矶时报》；*The New York Times*《纽约时报》；*The Washington Post*《华盛顿邮报》；*The Wall Street Journal*《华尔街日报》；*The New York Daily News*《纽约每日新闻》；*Chicago Daily Tribune*《芝加哥论坛报》；*USA Today*《今日美国》；*New York Post*《纽约邮报》；*The Christian Science Monitor*《基督教科学箴言报》；*International Herald Tribune*《国际先驱论坛报》；等等。

美国主要杂志的翻译：

Reader's Digest《读者文摘》；*TIME*《时代周刊》；*Life*《生活》；*People*《人民》；*Cosmopolitan women*《世界妇女》；*American Home*《美国家庭》；*American Child*《美国儿童》；*American Literature*《美国文学》；*Scientific American*《科学美国人》；*Playboy*《花花公子》；等等。

（2）音译。

E. g. （新加坡）联合早报 *LianheZaobao*；（香港）大公报 *Ta Kung Pao*；（香港）文汇报 *Wen Wei Po*；等等。

（3）借用原文的方式.

E. g. （俄罗斯）真理报 *Pravda*；（日本）朝日新闻 *The AsahiShimbun*；（法国）世界报 *Le Monde*；等等。

2. 文献、著作名的翻译

当行文涉及外国文献、著作的名称时,汉语文章中使用汉语译名,英语文章中使用英语译名、原名称的音译或罗马化拼写形式。

E. g. *The Wealth of Nations*《国富论》；*The Declaration of Independence*《独立宣言》；*The Origin of Species*《物种起源》；《红高粱》*Red Sorghum*；《红楼梦》*A Dream in Red Mansions*；《三国演义》*Romance of the Three Kingdoms*；《西游记》*The Journey to the West*；等等。

今天,汉语科技论文中的英语参考文献一般直接以英语原文的形式列出,而英语论文中引用的汉语参考文献,由于语体不兼容原因,必须罗马化,一般以英语译文的形式出现,有时以汉语拼音的形式出现。

5.2 科技英语术语的译法

术语是准确标志科学技术和社会领域一定概念的词语,是反映科学技术和社科领域进步的特殊标记,用来记录和表述各种现象、过程、特性、关系、状态等不同名称(方梦之,2011)。我国著名科学家钱三强曾说过:"科技名词术语是科学概念的语言符号。人类在推动科学技术向前发展的历史长河中,同时产生和发展了各种科技名词术语,作为思想和认识交流的工具,进而推动科学技术的发展。"由此可见术语的重要性。

当今,现代科技多个领域中科技的发明和命名工作都是西方发达国家完成的。科技英语文本中有很多术语,对这些术语的翻译极大地丰富了汉语语言,掌握这些术语对科技翻译帮助很大。

5.2.1 科技英语术语的构成

一般来说,科技英语术语由以下方式构成。

1. 缀合法

缀合法,即将前缀/后缀词和词根合成新的词汇的构词方法。

(1)前缀法。前缀法可改变词的意思,但是一般不会改变词性。

前缀可以用来表示多个含义。

1)表示时间顺序:往往以 fore-,post-,pre-,proto-,re-/retro-等开头。

E.g. foreground 前景;postnuptial 结婚后的;preatomic 原子能利用以前的;protozoology 原生动物学;reentry 重新进入;retrospective 回顾的,怀旧的;等等。

2)表示数目:往往以 mono-、uni-、semi-、bi-、di-、tri-、kilo-、centi-等开头。

E.g. monochrome 单色,单色画;nicameral 单院的,一院制的;semiannual 一年两次的,半年一次的;biannual 一年两次的;dichotomy 二分法,两分,分裂;kilogram 千克,公斤;centimeter 厘米,公分;等等。

3)表示否定:往往以 il-/ir-/im-/in-,a-/an-,un-,de-,dis-,non-等开头。

E.g. illegal 非法的;irregular 不规则的;imbalance 不平衡;inaccurate 不精确的,有错误的;anaerobic 厌氧的,无氧的;abnormal 不正常的,变态的,畸形的;unarmed 不带武器的,徒手的;deactivate 使(仪器)停止工作,使(化学过程)灭活化(或减活化、钝化);disinfect 给……消毒;non-biodegradable(物质或化学品)不可生物降解的;等等。

4)表示程度:往往以 hyper-、mini-、out-、over-、sub-、super-、sur-、ultra-、under-等开头。

E.g. hyperacid 酸过多的;hypertension 高血压;miniaturization 小型化,微型化;outclass 远高于,远胜于;subgenus 亚属;superabundant 过多的,大量的;surfeit 过度;ultrafashionable 极其流行的;underbred 没教养的,下流的,劣种的;等等。

5)表示位置、方向:以 e-/ex-/extra-、endo-、fore-、hyper-、hypo-、im-/in-、inter-、intra-/intro-、over-、sub-、trans-等开头。

E.g. hyperglycemia 血糖过高症;extrauterine 子宫外的;endoscope 内窥镜,内诊镜;forebrain 前脑;hypersonic 特超声速的;intercolonial 殖民地间的;intracardiac 心脏内

的;sub-irrigation 地下灌溉;transmarine 海外的,横越海洋的;等等。

　　6)表示部分与整体:往往以 omni-,quasi-,pan-,per-,para-等开头。

E.g. omnidirectional 全方向的;quasi-continuum 准连续区;pantology 百科全论,人类知识综合体系;paralinguistics 辅助语言学,派生语言学;等等。

　　7)表示异同:往往以 homo-,iso-,sym-/syn-,hetero-等开头。

E.g. homogeneity 同质,同种;isocline 等斜线;symmetry 对称(性),整齐,匀称;heterochromosome 异染色体,性染色体;等等。

(2)后缀法。后缀法会改变词性,但新构词与原词意思相近。

常见的后缀有:

1)形容词+后缀-ize/ise,-en 等,构成动词。

E.g. internationalize 使国际化;blacken 使变黑,诽谤,诋毁;等等。

2)名词+后缀-age,-ful,-ing,-ism,-ship,-eer,-er,-ess 等。

E.g. percentage 百分比,百分数;mouthful 一口,满口;paneling 嵌板;flagship 旗舰;engineer 工程师;Darwinism 达尔文主义,进化论,达尔文学说;conductress(公共汽车、电车等的)女售票员,【美国英语】(火车的)女列车长;等等。

3)名词+后缀-al,-ed,-esque,-ic,-ful,-like,-less,-ly,-ous,-y、-proof 等,构成形容词。

E.g. structural 结构上的;aged 年老的;picturesque 风景如画的;painless 无痛的,不痛的;metallike 金属般的;yearly 每年,一年一次;marvelous 了不起的,非凡的;waterproof 防水的;等等。

4)动词+后缀-ant,-er/or,-ee,-age,-al,-ation,-ing,-ment 等,构成名词。

E.g. assistant 助手,助理;computer 计算机;testee 应考人,测验对象;exploration 探测,探究;forging 锻造,锻件,伪造;development 发展,开发,发育;等等。

5)形容词+后缀-ity,ness,-cy 等,构成名词。

E.g. plasticity 塑性,柔软性;elasticity 弹性,弹力,灵活性;flexibility 灵活性,弹性;exactness 精确;adequacy 足够,适当;等等。

6)动词+后缀 able/-ible,-ful 等,构成形容词。

E.g. accountable 有责任的,可解释的;visible 看得见的;respectful 恭敬的;等等。

2.复合法

复合法是指将两个或两个以上的词合成一个新词的构词方法。

(1)复合动词。

E.g. mass-produce 大量生产;outsmart 比……更聪明,用计谋打败;co-work 协同工作;dry-clean 干洗;test-drive 试驾;等等。

(2)复合形容词。

E.g. airtight 密闭的;poverty-stricken 为贫穷所困扰的,非常贫穷;outstanding 杰出的,显著的;waterproof 防水的;snow-white 雪白的;white-hot 白热的,狂热的;painstaking 艰苦的,勤勉的;thought-provoking 发人深思的;等等。

(3) 复合名词。

E.g. keyboard 键盘;aircraft 飞机,航空器;bookcase 书柜,书架;bedrock 基岩;earthquake 地震,大动荡;landmark 陆标,地标;fieldwork 野外工作,现场工作,野战工事;cottonwood 杨木;seabeach 海滩;newsletter 时事通讯;等等。

3. 首字母缩略法

首字母缩略法,即将一连串词的首字母合成新词的方法,多用于机构名。

E.g. CIA = Central Intelligence Agency 中央情报局;FBI = Federal Bureau of Investigation(美国)联邦调查局;CPC = Communist Party of China 中国共产党;CPPCC = Chinese People's Political Consultative Conference 中国人民政治协商会议;LED = Light-Emitting Diode 发光二极管;SCI = Science Citation Index 科学引文索引;AIDS = Acquired Immune Deficiency Syndrome 艾滋病;等等。

4. 缩合法

缩合法是指将两个词各取一部分后合成新词的方法。

E.g. smog = smoke + fog 烟雾;medicare = medical + care 医疗保健;nanotech = nanometer + technology 纳米技术;telex = teleprinter + exchange 电传;telecast = television + broadcast 电视广播;motel = motor + hotel 汽车旅馆;Brexit = British + exit(out of Europe)英国脱欧;等等。

5.2.2　科技英语术语的翻译方法

1. 直译法

此处,直译法指的是将构成术语的词义拼接起来当作术语用。科技汉语大量的术语就是这样来的。

E.g. abdominal belt 腹带;absolute counter 绝对计数器;binary fission 二分裂;blue whale 蓝鲸;bombarding particle 轰击粒;companion plants 伴生植物;compensation circuit 补偿电路;complete reflection 全反射;continental deposit 大陆沉积;fish glue 鱼胶;等等。

2. 音译法

由于现代科技很多是西方发达国家发明和命名的,汉语中没有对应词,所以译入汉语时多采用音译法。

E.g. ampere 安培(计算电流强度的标准单位);calorie 卡路里(热量单位);ton 吨;hertz 赫兹(频率单位);nanometer 纳米(即十亿分之一米);Parkinson 帕金森(病);Einstein 爱因斯坦(著名物理学家);等等。

3. 音译+增译法

E.g. *The Times*《泰晤士报》;domino 多米诺骨牌;Citroen 雪铁龙轿车;sauna 桑拿浴;ballet 芭蕾舞;golf 高尔夫球;等等。(增译表功能的词,如"报""牌""车""浴""舞""球")

4. 形译法

形译法指的是根据所指物的实际形状翻译的方法。

E.g. H-beam 工字梁;O-ring 环形圈;V-belt 三角皮带;twist drill 麻花钻;U-steel 槽钢;

X-brace 交叉支撑;等等。

5. 归化法

归化法旨在尽量减少译文中的异国情调,为目的语读者提供一种自然流畅的译文。

E.g. diesel engine 柴油机;meter 米,仪表;decimeter 公寸,分米;transformer【电】变压器; autocoder 自动编码器;micromotor 微型电动机;等等。

练 习 题

一、请将下列文本译为汉语。

Five years ago, Michael Preysman swore that his online clothing company, Everlane, would never have a bricks-and-mortar store.

"We are going to shut the company down before we go to physical retail," he told *the New York Times*. He doubled down on that pledge in March, telling Quart z he couldn't think of any apparel stores that offer "a great experience."

Now, it seems, his view on that has, well, evolved.

Everlane, the socially minded brand Preysman founded six years ago, is opening its first two stores Dec. 2. The flagships, on New York's Prince Street and in San Francisco's Mission District, will carry many of the company's best sellers, including T-shirts, cashmere, denim and shoes. And although the stores will be relatively small — 2,000 square feet in New York, 3,000 square feet in San Francisco — Preysman says they will allow the company to reach new shoppers and interact more closely with existing ones.

"Our customers tell us all the time that they want to touch a product before they buy it," Preysman said. "We realized we need to have stores if we're going to grow on a national and global scale."

There were other reasons, too, he said: It turns out that even people who buy online prefer to return or exchange items in person.

Everlane, which Preysman founded at 25, is the latest online darling to set up physical locations. Warby Parker, the popular eyeglasses retailer, opened its first retail shop in New York in 2013. Today, it has more than 60 stores across the United States and Canada. Bonobos, the men's retailer that was bought by Walmart this year for $310 million, has expanded offline in recent years, as have shoe company M. Gemi and clothing brand Cuyana.

The trend underscores how quickly attitudes about technology can shift among retailers. Not long ago, some thought of bricks-and-mortar stores as vestiges of the past that add to operating expenses. But they also prove important in attracting new customers and fulfilling online orders. Even Amazon.com made its way into the bricks-and-mortar business this year with its $13.7 billion purchase of Whole Foods Market. (Amazon chief executive Jeffrey P. Bezos owns *The Washington Post*.)

"After you've gotten all the low-hanging fruit — customers who are willing to buy online or who follow your brand on social media — how else do you get people to buy your items?" said Sucharita Mulpuru, a retail analyst for market research firm Forrester. "That's where having a physical footprint can make an impact."

And, she added, the conditions for opening small-format physical locations have improved in recent years, as shopping centers look to fill vacancies. Given the widespread closures around the country — more than 7,000 stores have shuttered so far this year — landlords and shopping centers are more willing to forge flexible arrangements, including monthly or yearly leases, with niche brands.

(https://www.washingtonpost.com/business/economy/everlane-is-opening-its-first-stores-after-years-of-swearing-it-wouldnt/2017/11/22/32e7d142-c9ba-11e7-b0cf-7689a9f2d84e_story.html?utm_term=.f3c8d303be39)

二、请将下列文本译为英语。

台湾媒体称,中国海军"辽宁"号航母编队25日穿越宫古海峡时遭到日本潜舰企图混入编队中,随后航母护卫舰立刻出动直-9反潜侦察机投掷声纳追击,一路进逼至宫古岛10公里处。

日本电视台援引日本幕僚统合监的消息称,中国一架从航母护卫舰上起飞的直-9直升机在经过宫古海峡时起飞,在距离宫古岛西南方向10km至30km处空域进行飞行,但没有侵犯日本领空。日本航空自卫队紧急出动战斗机予以应对。

台媒认为,出动直-9直升机或有检测拦截日本潜水艇的可能。

日本防卫省此前说,派出了"五月雨"号护卫舰和那霸军事基地的P3C反潜警戒机对"辽宁"号航母编队监视跟踪,但并没有说日本潜艇跟踪监视的消息。

(http://news.qq.com/a/20161227/014578.htm)

第6章 科技文本中数词的翻译

表示数目的多少与顺序先后的词叫数词。在科技文章中,数词用来罗列顺序、计量、编码、明晰事物之间的数量关系,扮演着重要角色。

数字含义十分严格,翻译时力求准确,不容半点偏差。所谓"差之毫厘,谬以千里",在科技文本数词的翻译中表现得尤为明显,所以翻译时要特别注意。

数字可分为整数、概数、小数、分数和百分数。整数又分为基数和序数。基数词表示数目,如one,two,three,four等;序数词表示顺序,如first,second,third,fourth等。序数词应用时词前一般要加"the"。

英语和汉语的数字在进位上有差异。两者的进位都是从右到左,即从个位开始,汉语按照独位计数,如个、十、百、千、万等,相邻之间差10倍;英语按照每三位进位,用逗号隔开,从右开始,第一个逗号为"千"(thousand),第二个逗号为"百万"(million),第三个逗号前为"十亿"【billion(美式)或 thousand million(英式)】,第四个逗号前为"兆"【trillion(美式)】(傅勇林,唐跃勤,2012)。汉语和英语计数方式有所不同,所以翻译时要小心谨慎,避免错误。

6.1 确数的译法

确数,即表示准确概念的数,有时用阿拉伯数字书写,有时用汉字书写。

6.1.1 用阿拉伯数字表示的确数

在科技英语中,用阿拉伯数字表示确数是最为常用的表示方式,其译法如下。

1. 直译(原数照抄)

此时,数目往往在0~10万以内,用以记录年份、温度、价格、距离、金额、尺寸、产量、功率、压力、比率等。

E.g. at −460℃ 在−460℃;by 1999 到1999年(添加"年");

1,400 km 1 400 千米;656 mm×245 mm 656 毫米×245 毫米

2. 转换法

英汉计数单位差别很大,对于10万以上这样较大数字的翻译常常需要进行换算,英译汉时往往要将阿拉伯数字换算为汉语常用的"万""亿"等单位表达法。

E.g. 600,000,000 kilowatts 6亿千瓦;$6.3×10^9$ kcal 6.3亿千卡;

23,367,000 tons 2 336.7万吨;100 million cases of diseases 1亿病例

3. 译为汉字

当较大的数字为"万""亿"的整倍数时,可将数词译成汉字。

E.g. 9,600,000 square kilometers 九百六十万平方千米;1,000,000 dollars 一百万美元;
65,123,000 cubic meters 六千伍佰一十二万三千立方米;300,000,000 tons 三亿吨

6.1.2 用文字表示的确数

对于英语中用文字数词或阿拉伯数字表示的数量,可根据具体情况译为阿拉伯数字或汉语。

E.g. 5 hundred thousand yuan 50 万元/五十万元;
one million barrels of oil 100 万桶石油/一百万桶石油

银行系统为确保资金金额明确、不易涂改,往往将金额全部用大写汉字表达出来,即用"零、壹、贰、叁、肆、伍、陆、柒、捌、玖、拾"这些汉字表达金额。

E.g. Twenty-Two Thousand Three Hundred and Fifty Pounds Only
贰万贰千叁佰伍拾英镑整

6.2 概数的译法

概数,即大概的数目,一般用"几""一些""以下""余""成千""上万""左右""上下"等词语表示,或用数词连用的方式表示,如"三四个""五六天""七八百人"等。

6.2.1 采用整数复数

E.g. tens of... 数十/几十;decades of... 数十/几十;dozens of... 几十/许多;scores of... 许多/大量(40~100 多);hundreds of... 几百/数百/成百上千的;thousands of... 几千/成千上万的;tens of thousands of... 几万/数万;hundreds of thousands of... 几十万/数十万;millions of... 几百万/千千万万的/数百万;tens of millions of... 几千万/数千万;millions and millions of... 亿万;等等。

6.2.2 采用习惯表达法

E.g. ten to one 十之八九/十有八九;the nineties 90 年代;the early 1860's 19 世纪 60 年代初期;in her sixties 在她六十多岁时;down to 300 kg/t ≥300 千克/吨;close to one thousand people 近一千人;within a factor of 在……范围内;等等。

6.2.3 其他情形

数词与一些词/短语相用,也可以表示概数。这些词常见的有"odd"(多)、"more than"(超过)、"over"(在……以上)、"above"(在……以上;超过)、"up to"(多达)、"less than"(少于)、"under"(在……以下)、"below"(在……以下)、"some"(大约)、"about"(大约)、"approximately"(大约;接近)等。

E.g. twenty odd 二十几/二十多/二十有余;two hundred or odd 二百多;more than eighty 八十多/八十以上/多于八十;up to 20 years 多达 20 年;less than five hundred 五百以下/不到五百/少于五百;under/below four thousand 四千以下/不到四千/少于四千;等等。

E.g. Although the company generated more than $1.3 billion in revenues in 2014, it spent less than $2 million in advertising.

虽然该公司在2014年收入超过13亿美元，其用于广告的费用却不足200万美元。

6.3 增减数的译法

在英语中，数的增减往往借助表示数量增减的词汇来表达，表达"增加""减少""降低""下降"等意思。这些词汇包括increase, rise, raise, grow, go up, decrease, diminish, fall, reduce, lower, drop, go down等。

E.g. The yield of apples in this region has increased by 2 times as compared with that of 2017.

与2017年相比，该地区苹果产量增长了2倍。

E.g. 1960年，美国上升，加拿大下降，两国数据分别是118 000例和102 000例。

In 1960, U.S. rose while Canada fell and the figures for both countries were 118,000 cases and 102,000 cases respectively.

6.4 倍数的译法

6.4.1 倍数的比较和增加

常见的倍数比较有以下4种句型，一般译为"A是B的N倍（大小、长度、质量等方面）"，或"A比B大N－1倍（大小、长度、质量等方面）"。

(1) A is N times as large (long, heavy...) as B.

(2) A is N times larger (longer, heavier...) than B. （注意：不是大N倍，而是N－1倍）

(3) A is larger (longer, heavier...) than B by N times.

(4) A is N times the size (length, amount,...) of B.

在表达倍数增加时，英汉两种语言恰好相差一倍。汉语中的"增加了N倍"，即表示纯增加的倍数，在译成汉语时应将英语的倍数减"1"，如可译"增加了N－1倍"，或"增加到N倍/是原有的N倍"。

常见的句型有以下两种。

(1) increase N times; increase by N times; increase by a factor of N; increase N fold
"增加了N－1倍"或"增加到N倍"。

(2) increase to N times "增加到N倍"或"增加了N－1倍"。

E.g. Thanks to the mechanization, the speed of harvesting crops has increased ten-fold.

多亏了机械化，收割庄稼的速度比以往提高了9倍（或"是以往的10倍"）。

6.4.2 倍数的减少

常见的表示倍数减少的句型有以下两种。

(1) decrease/drop/reduce（to）＋ N times；N-fold reduction；N times less than；reduce by a factor of N 都指"减少了 N 分之 N－1 倍"或"减少到 N 分之 1 倍"

与英语不同，汉语一般不说减少了 N 倍，而说"减少了几分之几"或"减少到几分之几"，即"减少了 N 分之 N－1"或"减少到减少到 N 分之 1"，如 reduce 5 times 可译成"减少了五分之四"或"减少到五分之一"。

E. g. Switching time of the new-type transistor is shortened 3 times.

新型晶体管的开关时间缩短了三分之二（即减少到原来的三分之一）。

E. g. The equipment reduced the error probability by a factor of 5.

该设备的误差概率降低了五分之四。

如果减少的倍数中有小数点，应换算成不带小数点的分数。例如，reduce 5.5 times 应译成"减少了十一分之九"或"减少到十一分之二"。

E. g. The plastic container is 4.5 times lighter than that glass one.

这个塑料容器比那个玻璃容器轻九分之七（即质量仅为原来的九分之二）。

(2) half as much as 和 twice less than 都指"比……少一半"或"比……少二分之一"。

E. g. South koran households today save only half as much as 10 years ago.

今天，韩国家庭储蓄的钱仅为十年前的一半。

练 习 题

一、请将下列文本译为汉语。

In a brief speech at the Jiuquan Satellite Launch Center, Chinese Premier Wen Jiabao hailed the successful launch of Shenzhou-6, will be recorded in the country's glorious history. China's space program, which is purely for peaceful purposes, is a contribution to human's science and peace, Wen said, adding that China is willing to cooperate with other nations in the development of space science and technology. China has been implementing a three-stage manned space flight program, culminating in the establishment of a permanent space laboratory.

Shenzhou Ⅵ astronauts performed more space tests

Two astronauts aboard China's second manned space flight moved between the re-entry and orbital modules of theShenzhou Ⅵ spacecraft on Thursday, as they continued tests designed to pave the way for an ambitious program of space walks, docking manoeuvres and a space laboratory. The transfer between the modules mainly tested the impact of movement on the spacecraft, the government's official Xinhua news agency said. Fei Junlong left the astronauts'space capsule and entered the orbital module for the first time on Wednesday. Fei's colleague, Nie Haisheng, performed similar work on Thursday on equipment mounted in the orbital module. The carrying of two astronauts and the movement into the orbital module marks new progress in China's space program.

Successful Landing

The re-entry capsule of China's second manned spacecraft landed safely early Monday in the remote Inner Mongolia region, after a five-day mission that was designed to push forward Beijing's ambitious space program. The Shenzhou Ⅵ capsule landed at 4:32 a.m. (20:32 GMT Sunday), bringing astronauts Fei Junlong and Nie Haisheng back to heroes' welcomes after a flight lasting 115 and a half hours. Wu Bangguo, in the hierarchy of China's ruling Communist Party, said the success of the mission was "of great significance for elevating China's prestige in the world and promoting China's economic, scientific and national defense capabilities, and its national cohesiveness". The government spent 900 million yuan (110 million dollars) on Shenzhou Ⅵ, Tang Xianming, director of China Manned Space Engineering Office, told reporters. Chen Bingde, the top military commander overseeing Shenzhou Ⅵ, declared the mission a "complete success". Chen and Defence Minister Cao Gangchuan led a column of military leaders who welcomed Fei and Nie when they arrived by plane in Beijing, where their families also met them. The astronauts, both People's Liberation Army colonels and former fighter pilots, said they were "in good condition", the government's official Xinhua news agency said. Fei climbed out of the capsule unaided, followed by Nie, the agency said. Initial medical checks showed the astronauts had "normal physical indications". The landing was about three hours earlier than originally planned. Before reentry, the astronauts separated their reentry capsule from the orbital capsule, the "nose" of the spacecraft that can remain in orbit for up to eight months as an electronic and optical surveillance satellite.

(秦荻辉. 精选科技英语阅读教程[M]. 西安:西安电子科技大学出版社,2008.)

二、请将下列文本译为英语。

特点

计算机系统的特点是能进行精确、快速的计算和判断,而且通用性好,使用容易,还能联成网络。①计算:一切复杂的计算,几乎都可用计算机通过算术运算和逻辑运算来实现。②判断:计算机有判别不同情况、选择作不同处理的能力,故可用于管理、控制、对抗、决策、推理等领域。③存储:计算机能存储巨量信息。④精确:只要字长足够,计算精度理论上不受限制。⑤快速:计算机一次操作所需时间已小到以纳秒计。⑥通用:计算机是可编程的,不同程序可实现不同的应用。⑦易用:丰富的高性能软件及智能化的人-机接口,大大方便了使用。⑧连网:多个计算机系统能超越地理界限,借助通信网络,共享远程信息与软件资源。

组成

图1为计算机系统的层次结构。内核是硬件系统,是进行信息处理的实际物理装置。最外层是使用计算机的人,即用户。人与硬件系统之间的接口界面是软件系统,它大致可分为系统软件、支援软件和应用软件三层。

硬件

硬件系统主要由中央处理器、存储器、输入输出控制系统和各种外部设备组成。中央处理器是对信息进行高速运算处理的主要部件,其处理速度可达每秒几亿次以上操作。存储器用于存储程序、数据和文件,常由快速的主存储器(容量可达数百兆字节,甚至数吉字节)和慢速海量辅助存储器(容量可达数十吉或数百吉以上)组成。各种输入输出外部设备是人机间的信息转换器,由输入-输出控制系统管理外部设备与主存储器(中央处理器)之间的信息交换。

软件

软件分为系统软件、支撑软件和应用软件。系统软件由操作系统、实用程序、编译程序等组成。操作系统实施对各种软硬件资源的管理控制。实用程序是为方便用户所设,如文本编辑等。编译程序的功能是把用户用汇编语言或某种高级语言所编写的程序,翻译成机器可执行的机器语言程序。支撑软件有接口软件、工具软件、环境数据库等,它能支持用机的环境,提供软件研制工具。支撑软件也可认为是系统软件的一部分。应用软件是用户按其需要自行编写的专用程序,它借助系统软件和支援软件来运行,是软件系统的最外层。

(https://baike.so.com/doc/5912132-6125040.html)

第7章　科技汉语无主句的翻译

汉语无主句指的是没有主语的汉语句子，是汉语常见的句式。无主句只说明事实本身，不明确说明主语，这在强调客观事实的科技文本中尤为常见。英语句子一般都必须有主语和谓语成分，句子结构才算完整。英语和汉语在句式建构上有差别。

7.1　英汉句式建构差异

英语和汉语属于不同语系，在形态和句式架构上有很大差别。

英语属于印欧语系。英语句子（尤其是长句）有较多修饰语，这些修饰语还可以有其自己的修饰语，句子不断延伸。句式结构虽然冗长复杂，但是因为逻辑关系明示，所以主次分明，条理清晰。

汉语属于汉藏语系。汉语句子往往按照自然事理的发展顺序和客观的因果关系展开，修饰语较短，句子由多个分句构成，分句与分句、短语与短语之间有语义联系，但是几乎不用关联词，主谓结构有时不明显，句子结构显得松散，短句多。

我国语言学家赵元任发现，英语属于"主语显著"的语言，汉语属于"话题显著"的语言。"主语显著"是指主语和谓语是句子结构的最基本语法单位，即英语句子至少有主语和谓语两个成分。英语句子多呈现"主语＋谓语（最重要信息）＋废话（其他相对不重要信息）"结构。"话题显著"是指句子往往先抛出一个话题，然后进行阐述或评价。句子结构呈现"话题和评论"模式，具体表现为"主语＋废话（其他相对不重要信息）＋谓语（最重要信息）"结构。

正是英语和汉语在句式建构上有很大差异，与英语句子相比，汉语中有大量的无主句。

E. g. 尚未决定何时交货。

It has not been settled yet when the goods will be delivered.

在这里，汉语的无主句转换成了英语中以形式主语"it"开头的句型。

E. g. 世界上已经建造了一些机器来从受控热聚变中发电。

Some machines have been built to produce electricity from controlled thermonuclear fusion.

在这里，汉语的无主句转换成了英语的被动语态。

7.2 科技汉语无主句的译法

科技文本中汉语无主句往往要采用以下方式译为规范的英语。

7.2.1 译为被动句

如果汉语无主句中包含动词和动作承受者（即宾语），翻译时即可将动作承受者译作译文英语的主语，将动作改为被动形式作谓语，进行翻译。

E.g. 自 2014 年 3 月将"大数据"首次**写入**《政府工作报告》以来，李克强总理在多个场合提及这一"热词"。

Since "big data" **was** first **written** in The Government Work Report in March, 2014, Premier Li Keqiang mentioned this "hot phrase" on several occasions.

在本句译文中，将原文中做逻辑宾语的"'大数据'"用作英语译文从句的主语，以被动语态重现原文信息，这样表达既客观又准确。

E.g. 昨天召开的常务会议**通过**《关于促进大数据发展的行动纲要》。

The Outlines of Action for Promoting the Development of Big Data **was passed** yesterday at the executive meeting.

在这里，原文中的逻辑宾语"《关于促进大数据发展的行动纲要》"转换成了译文的主语，译文采用被动语态结构。

7.2.2 使用动词短语翻译

如果无主句中的动词能译成动词短语，如准备（make preparation for）、规划（make planning）、注意（pay attention to）、结束（put an end to）、强调（lay emphasis on）、重视（attach importance to）、突出（give prominence to）、使用（make use of）、努力（make efforts to）、优先考虑（give priority to）、道歉（make an apology）等，翻译时可以将短语中的名词用作英语译文的主语，将句式变为被动结构。

E.g. 必须适当地**使用**国家的森林资源。

Proper **use** must be **made of** the forest resources of the country.

译文包含"make use of"这个短语，用"use"作译文主语，译文谓语变为被动形式。

又如：

E.g. 制定大数据行动纲要，要**突出**"政府大数据建设"和"创造健康发展的大数据环境"这两项核心内容。

Prominence must be **given** to two core contents, namely, "governmental construction of big data" and "creating the healthily-developing environment of big data", in order to enact the outlines of action.

这个译文包含"give prominence to"这个短语，用"prominence"作译文主语，译文谓语变为被动形式。

7.2.3 译为形式主语句、祈使句、存在句等特殊句型

当汉语无主句表示必然性和必要性时，往往译为英语的形式主语 it 引导的句子：It is

necessary /essential /imperative...。

E. g. **It is imperative** to strengthen the exploration and research on consular protection system.

加强对领事保护法律制度的探索与研究是**当务之急**。

E. g. **有必要**改进生产方法，提升产品质量。

It is necessary to improve production methods to raise the quality of products.

当汉语无主句表指令、命令、请求、号召等意思时，可译为英语的祈使句。

E. g. 在水中溶解10克氢氧化钠，稀释至1升。

Dissolve 10g of sodium hydroxide in water and dilute to 1 L.（祈使句）

无主句表示事物存在、出现或消失的，可以采用英语的存在句翻译。

E. g. 在某些型号的车床上出现取消尾架的趋向。

There is a trend to eliminate the tailstock on some models of the lathe.（存在句）

7.2.4　译为常见的被动句型

英语中有一类以 it 作为形式主语的被动句较为常见。常见的句型有：

It is said that... 据说……；It is well known that... 众所周知……；It is acknowledged that... 众所周知……；It is reported that... 据报道……；It is estimated that... 据估计……；It can be assumed that... 假定……；It can be expected that... 可以预期的是……；It is presumed that... 一般认为……；It must be admitted that... 必须承认……；It can be anticipated that... 可以预计……；It is believed that... 我们（人们）相信……；It must be pointed out/made clear that... 需要指出……；等等。

E. g. 据说，世界上运转着数以千计的各种地震仪器日夜监测着地震的动向。

It is said that thousands of kinds of seismic instruments in operation around the world monitor the trend of earthquakes day and night.

原文中"据说"对应英语的"It is said that..."，一下子就搭建起了英语译文的句子框架。

E. g. 需要指出的是，对于地震，我们更应该做的是提高建筑抗震等级、做好防御，而不是预测地震。

It must be pointed out/made clear that what we must do more is improve the seismic grade of buildings with an earnest defense rather than make earthquake prediction.

原文中"需要指出的是"被恰当地转换为"It must be pointed out/made clear that..."，引出后边的主要信息。

这类固定句型可以简化。具体做法是：把 it is presumed, it is said, it is thought 等句型中的动词 be presumed, be said, be thought 等放回后面从句的主谓之间。

E. g. 一般认为元宵节起源于汉朝。

The Lantern Festival **is generally thought** to have originated in the Han Dynasty.

翻译"一般认为"时，如果套用固定句型，即是"It is generally thought that..."，所以原文也可以翻译成：It is generally thought that the Lantern Festival originated in the Han Dynasty.

练 习 题

一、请将下列文本译为汉语。

Network topology

The physical layout of a network is usually less important than the topology that connects network nodes. Most diagrams that describe a physical network are therefore topological, rather than geographic. The symbols on these diagrams usually denote network links and network nodes.

Network links

The transmission media (often referred to in the literature as the physical media) used to link devices to form a computer network include electrical cable (Ethernet, HomePNA, power line communication, G. hn), optical fiber (fiber-optic communication), and radio waves (wireless networking). In the OSI model, these are defined at layers 1 and 2 — the physical layer and the data link layer.

A widely adopted family of transmission media used in local area network (LAN) technology is collectively known as Ethernet. The media and protocol standards that enable communication between networked devices over Ethernet are defined by IEEE 802. 3. Ethernet transmits data over both copper and fiber cables. Wireless LAN standards (e. g. those defined by IEEE 802. 11) use radio waves, or others use infrared signals as a transmission medium. Power line communication uses a building's power cabling to transmit data.

Wired technologies

Bundle of glass threads with light emitting from the ends.

Fiber optic cables are used to transmit light from one computer/network node to another.

The orders of the following wired technologies are, roughly, from slowest to fastest transmission speed.

Coaxial cable is widely used for cable television systems, office buildings, and other work-sites for local area networks. The cables consist of copper or aluminum wire surrounded by an insulating layer (typically a flexible material with a high dielectric constant), which itself is surrounded by a conductive layer. The insulation helps minimize interference and distortion. Transmission speed ranges from 200 million bits per second to more than 500 million bits per second.

ITU-T G. hn technology uses existing home wiring (coaxial cable, phone lines and power lines) to create a high-speed (up to 1 Gigabit/s) local area network.

Twisted pair wire is the most widely used medium for all telecommunication. Twisted-pair cabling consist of copper wires that are twisted into pairs. Ordinary telephone wires consist of two insulated copper wires twisted into pairs. Computer network cabling (wired Ethernet as defined by IEEE 802.3) consists of 4 pairs of copper cabling that can be utilized for both voice and data transmission. The use of two wires twisted together helps to reduce crosstalk and electromagnetic induction. The transmission speed ranges from 2 million bits per second to 10 billion bits per second. Twisted pair cabling comes in two forms: unshielded twisted pair (UTP) and shielded twisted-pair (STP). Each form comes in several category ratings, designed for use in various scenarios.

An optical fiber is a glass fiber. It carries pulses of light that represent data. Some advantages of optical fibers over metal wires are very low transmission loss and immunity from electrical interference. Optical fibers can simultaneously carry multiple wavelengths of light, which greatly increases the rate that data can be sent, and helps enable data rates of up to trillions of bits per second. Optic fibers can be used for long runs of cable carrying very high data rates, and are used for undersea cables to interconnect continents.

Price is a main factor distinguishing wired- and wireless-technology options in a business. Wireless options command a price premium that can make purchasing wired computers, printers and other devices a financial benefit. Before making the decision to purchase hard-wired technology products, a review of the restrictions and limitations of the selections is necessary. Business and employee needs may override any cost considerations.

(https://en.wikipedia.org/wiki/Computer_network)

二、请将下列文本译为英语。

文本1：

游客须知

为了使广大游人能在一个环境优美、秩序井然、清洁卫生、服务优质的环境下游玩、散步、休闲和娱乐，请自觉遵守以下规定：

1. 严格遵守我市颁发的"七不准"公告。
2. 不准自行车和各种机动车入园。
3. 不准携带各种宠物入园。
4. 服从管理、讲卫生、遵守社会公德。
5. 不准在园内捕鸟。

（来自百度图片内容）

文本2：

长江发源于"世界屋脊"——青藏高原的唐古拉山脉各拉丹冬峰西南侧。干流流经青海、西藏、四川、云南、重庆、湖北、湖南、江西、安徽、江苏、上海11个省、自治区、直辖市，于崇明岛以东注入东海，全长约6 300 km，比黄河（5 464 km）长800余千米，在世界大河中长度仅次于非洲的尼罗河和南美洲的亚马逊河，居世界第三位。但尼罗河流域跨非洲9国，亚马逊河流域跨南美洲7国，长江则为中国所独有。

(https://baike.baidu.com/item/%E9%95%BF%E6%B1%9F/388)

第8章　科技文本中被动语态的翻译

英语的语态有主动语态和被动语态共两种，用以说明主语与谓语之间的关系。被动语态由"助动词 be＋及物动词的过去分词"构成，表示主语是谓语动作的承受者，可以客观地表达意思而不必指出动作的发出者。

科技信息注重事实，所以被动语态在科技英语中使用极其广泛。汉语中的被动逻辑很少用"被"字表达，更多时候用其他词替代，比如"为""于""见""遭到"等词。为了保证汉语译文读来顺畅，译者英译汉时须注意转换被动情形。

英译汉时，被动语态常见的有以下三种处理方法。

8.1　译为主动句

8.1.1　被动句型1："主语（受动者）＋谓语（被动形式）＋by＋施动者"

当被动句结构是："主语（受动者）＋谓语（被动形式）＋by＋施动者"时，翻译时有以下处理方法：

1. 按"施动者＋动作＋受动者"顺序翻译

当施动者、动作、受动者三个因素同时存在时，翻译会变得简单很多。

E. g.　The Internet itself is considered by many enterprises to be an inadequately secure medium for the transport of confidential company data.

许多企业认为，通过互联网传输公司的机密信息不够安全。

E. g.　A number of cars are already on the road powered by gas turbines.

燃气轮机驱动的若干汽车已投入使用。

这里，施动者（gas turbines 燃气轮机）、动作（power 驱动）和受动者（a number of cars 若干汽车）清晰地呈现在原文中，所以只需按"施动者＋动作＋受动者"顺序翻译即可。

E. g.　The Lavi fighter plane that was to have been developed by Israel depended heavily upon American funding and technology.

本该由以色列自主研发的幼狮战斗机对美国提供的资金和技术依赖十分严重。

原句中施动者（Israel）、动作（develop）和受动者（The Lavi fighter plane）都有，所以翻译起来就简单了。

2.将"by+施动者"译为状语

E.g. Without the cooling and lubrication systems, an engine would be badly damaged by heat and friction within minutes.

如果没有冷却和润滑系统,发动机将在几分钟内因为发热和摩擦而严重受损。

虽然施动者(heat and friction)、动作(damage)和受动者(an engine)都有,但是"因为发热和摩擦而严重受损"这个因果关系应该更符合汉语的表达习惯,所以我们把"by+施动者"翻译为状语。

E.g. It is reported that metal removal rates and surface finish are controlled by the frequency and intensity of the spark.

据报道,金属切削速度和表面光洁度是通过电火花的频率与强度来控制的。

在这里我们看到,"control"这个动作的发出者是"the frequency and intensity of the spark",连同介词"by"被译成了方式状语:"通过电火花的频率与强度",这样译文读来更流畅。

3.将原文信息结构调整为"主语(即原主语,受动者)+谓语(即原谓语,主动式)+宾语(原施动者)"结构

汉语是形合语言,信息之间的关系靠内在衔接,到底是主谓关系还是动宾关系要靠上下文来判断,所以英译汉时有时可以省略英语助动词"be"(被)。

E.g. As we have mentioned above, coal production and use is accompanied by serious environmental problems.

正如我们上面所提到的,煤炭的生产和使用伴随有严重的环境(污染)问题。

尽管"煤炭的生产和使用"和"伴随"是动宾关系,但是译文将两者处理成表面上的主谓关系,没有用"被"字,将英语明示的关系在汉语一定程度上"隐藏"掉了。

4.将原文信息结构调整为"主语(即原主语,受动者)+表被动意词+宾语(原施动者)+谓语(即原谓语,主动式)"结构,或"主语(即原主语,受动者)+表被动意词+谓语(即原谓语,主动式)+宾语(原施动者)"结构

汉语中很少用"被"字表被动意思,往往会换成其他词来表意,如"受""遭到""于"等词。

E.g. The complex process of photosynthesis is controlled by several gene.

光合作用过程复杂,受数种基因制约。

译文中用"受"字来表达被动含义。

E.g. Square waves are universally encountered in digital switching circuits and are naturally generated by binary (two-level) logic devices.

(电子)方波在数字开关电路中普遍存在,由二进制(二级)逻辑装置自然生成。

译文中用"由"字来表达被动含义。

E.g. Current is equal to voltage divided by resistance.

电流等于电压除以电阻。

译文中用"以"字表达出了被动含义。

E. g. It means that the speed of a falling object is not determined by its quality.

这表明,物体下落的速度不取决于其质量。

译文中用"于"字表达出了被动含义。

5. 其他处理方法

E. g. The computer is chiefly characterized by its accurate and quick computations.

这台计算机的主要特点是准确性高且速度快。

译文采用词类转换法,把原文中的动词"characterize"译为名词"特点",没有用"被"字。

8.1.2 被动句型2:"主语+谓语+状语"

这种结构译为主动句时,往往有以下处理方式:

1. 将原文信息结构调整为"状语+谓语+宾语(即原主语)"结构

E. g. An extract in marijuana that relieves pain and muscle spasms in MS patients has been approved in Europe and Canada.

欧洲和加拿大已经批准使用大麻提取物,以缓解疼痛和多发性硬化病人肌肉痉挛。

原文信息结构调整为"状语(即欧洲和加拿大)+谓语(批准)+宾语(原主语:大麻提取物)"结构,这样译文句式符合汉语的表达习惯。

以下2个例子是同样的处理办法:

E. g. Matter is transported in a circulatory system.

循环系统输送物质。

E. g. The deficiency of such observations will be eliminated by the installation of more ocean bottom seismographs.

安装更多的海底地震仪可以弥补这种观测上的不足。

2. 将原文信息结构调整为"主语(即原主语)+状语+谓语(即原谓语,主动式)"结构

E. g. When the histamines are released in the skin, you get a type of rash called hives.

当组胺在皮肤中释放时,就会引发一种皮疹,称作荨麻疹。

原文"histamine(组胺)"和"release(释放)"的动宾关系被处理成译文中表面上的主谓关系,没有用"被"字。

8.2 译为被动句

将英语的被动句译为汉语包含"被"字的句子虽然少见,但是的确存在。这样做往往是为了达到强调目的,突出受动者。

E. g. If any animals or persons were to be caught by the electric discharge of an electron beam weapon, they would most likely be electrocuted.

任何动物或人,如果被电子束武器的放电所击到的话,很有可能被电死。

译文中保留了"被"字。

E. g. If the scheme is disapproved, work on the project will stop immediately.

如果该方案被否决,这项工程将立即停工。

译文中保留了"被"字。这样做往往是为了强调,突出受动者。

8.3 按套路翻译

被动句型:It + be +过去分词。

这个句型汉译英有一定的套路。常见的句型有:

It is thought that... 人们认为……;It is said that... 据说……;It is reported that... 据报道……;It is estimated that... 据估计……;It is presumed that... 一般认为……;It must be admitted that... 必须承认……;It is believed that... 我们(人们)相信……;It is well known that... 众所周知……;It must be pointed out that... 必须指出……;It is regarded that... 人们认为……;It is stressed that... 人们强调……;It is asserted that... 有人主张……;It is asked that... 有人问……;等等。

E. g. It was once believed by scientists that xenon could not form chemical compounds.

科学家曾经相信,氙气不能形成化合物。

因为原文中有动作"believe"的发出者"scientists",所以译文直接处理成主谓关系,即"科学家相信……"。

E. g. It is widely acknowledged that these materials can withstand strong stress and high temperature.

普遍认为,这些材料能承受强大应力和高温。

这里是按照套路翻译的,将"It is widely acknowledged that"句式翻译为"普遍认为"。

E. g. It has long been known as a source of unusually strong radio signals.

人们早就知道那是一个异常强大的射电信号源。

汉语常从人的角度出发来叙述客观事物或描述人及其行为或状态,常常使用人称作句子主语,所以译文合理添加"知道"的发出者"人们",也可将这句视作按照套路转换语态的。

练 习 题

一、请将下列文本译为汉语。

Winter sun getaways: readers' travel tips

Winning tip: near Port Antonio, Jamaica

If you dream of sandy beaches and 30℃ sun this winter but your wallet is more accommodating of a weekend in Scarborough, how about staying in a local village in your dream destination? In the foothills of the Blue Mountains of Jamaica near Port Antonio is a charming cottage called Papaya. Accommodation is basic yet naively beautiful, easily forgiven when you are settling down to read your book in the veranda hammock. The only sounds you hear are woodpeckers, parrots and the leaves in the trees. There is no tourism here, no one wanting you to buy anything, just friendly local people getting on with their day with a greeting and smile. Papaya is a two-bedroom cottage, sleeping four, and will set you back £10 a night each in January — yes really!

Martin Colegate
Kotu, Gambia

Photograph: Alamy

Sipping a cold beer at a beach bar in the sun is my idea of a perfect winter holiday. The relaxed and beautiful Gambian coast at Kotu is cheaper than many winter sun destinations and accessible thanks to charter flights. The Kombo Beach Hotel has a range of bars and restaurants and a laidback vibe. Bed and breakfast for two costs around £70 a night in December. Liv bar, the most popular in the area, is within the complex so no need to pay for a taxi back from the best night-time drinking around.

Alvor, Portugal

If you could flee the British winter to a superbly appointed apartment with private roof terrace; if that apartment was within a few minutes' walk of theunspoilt, Moorish-influenced fishing village of Alvor and its gorgeous beach — you'd be tempted, yes? We were and paid the modest sum of £35 a night with Owners Direct. We felt part of a genuine Portuguese community here, shopping and eating (usually that morning's catch) with the locals. Lazy days were interspersed with walks along the boardwalk and estuary trail, exploring the glorious coast. Cheap flights to Faro make this an affordable winter escape.

PlayaLarga, Cuba

Our casa — privately owned guesthouse (£40 a night for a family room, via cuba-junky.com) — was right on a picture-perfect beach at PlayaLarga, Bay of Pigs. We saw flamingos, and went snorkelling in the sea and in the Cueva de losPeces. History and culture surround you. Cubans who lost their lives when the Americans invaded have a gravestone marking where they fell; the taxi we got back to Havana was a 1950s Buick; and the yellow American school bus used for the snorkelling dodged tyre-threatening crabs.

(https://www.theguardian.com/travel/2017/nov/23/winter-sun-getaways-readers-tips-algarve-thailand-caribbean)

二、请将下列文本译为英语。

2015年8月21日中国证监会新闻发布会汇总

2015年08月21日 20:04　　来源：中国经济网

　　2015年8月21日，证监会召开新闻发布会，新闻发言人张晓军通报了近期查处上市公司大股东及实际控制人违法案件的情况，向港澳地区扩大证券经营机构宣传了对外开放的有关政策（以上内容见官网要闻栏目），最后回答了记者的提问。

　　问：监管部门是否暂停了分级基金的审批工作？分级基金指引政策进展情况如何？

　　答：考虑到分级基金机制较为复杂，普通投资者不易理解，且前期出现了一些新情况、新问题，为此，我会暂缓了此类产品的注册工作，并正在研究有关政策。

　　问：《证券法》修订草案何时进行二审？目前进展情况如何？

　　答：《证券法》的修订是证券市场的一件大事，社会各界对此高度关注。今年4月，《证券法》修订草案已由第十二届全国人大常委会第十四次会议第一次审议。目前，立法机关正按照立法程序对修订草案进行研究完善。证监会将全力配合立法机关做好相关工作，研究并反馈立法需求与意见建议，希望尽快安排第二次审议。

责任编辑：魏京婷

(http://finance.ce.cn/rolling/201508/21/t20150821_6290828.shtml)

第 9 章　科技语篇翻译

9.1　语篇及其特征

语篇(discourse/text)指的是实际使用的语言单位,是交流过程中的一系列连续的语段或句子所构成的语言整体(360 百科)。在篇幅上,语篇是大于句子的语言单位,可以短到一则通知,长到系列小说;在功能上,语篇是一种独立、完整的交际行为。语篇内各成分之间在形式上是衔接(有形)的,在语义上是连贯(无形)的,句与句之间的排列符合逻辑,信息不断向前推进。

语篇有以下基本特征:
(1)语篇具有中心思想(controlling idea),这是语篇的灵魂所在;
(2)段落具有段落大意(the gist of a paragraph);
(3)信息具有整体性(unity)和一致性(consistency)(内容上,甚至风格上);
(4)信息具有连贯性(coherence)。

9.2　语篇常见的布局模式

一般情况下,语篇(以文章为例)呈现以下布局模式:

- 主题(中心思想)
Ⅰ.简介
Ⅱ.主要观点 1
　A.从属观点 1
　　1.阐明观点的细节
　　2.进一步拓展信息(手段:举例、作比较、打比方、列举、引用等)
　B.从属观点 2
　　1.阐明观点的细节
　　2.进一步拓展信息(手段同上)
Ⅲ.主要观点 2
　A.从属观点 1

1. 阐明观点的细节
2. 进一步拓展信息(手段同上)
B. 从属观点2
1. 阐明观点的细节
2. 进一步拓展信息(手段同上)
C. 从属观点3
Ⅵ. 结论

(比如)以下是一篇文章的布局模式:

- 主题(中心思想):百慕大三角区轮船、飞机失事/失踪有三个原因。

Ⅰ. 简介:每年在该地区依然有轮船、飞机失事/失踪等神秘事件发生。
Ⅱ. 这跟恶劣天气有关。
  A. 该地区地理位置特殊,天气多变;
  B. 该地区常有飓风、雷暴雨和龙卷风天气;
     天气会突变,猝不及防。
     a. 实例1:该地区海浪高达4.6~6米;
     b. 实例2:"巴尔的摩荣光号"船沉没。
Ⅲ. 这也跟海面不断溢出的沼气有关。
   巨大的沼气泡猛烈喷出,令浮力减小或消失。
Ⅳ. 这也跟神秘的时光隧道有关。
   时光隧道令时间紊乱。
Ⅵ. 结论:百慕大三角区轮船、飞机失事/失踪等神秘事件原因探究仅处于理论层面,还没有定论。

假如翻译这样一篇文章,理论上讲,译者提供的译文也应该是这样"总-分-总"模式,只是语言是目的语。实际上,这是很多段落/文章呈现的模式。

E. g. One of the more widely publicized dangers is that of Legionnaire's disease, which was first recognized in the 1970s. This was found to have affected people in buildings with air-conditioning systems in which warm air pumped out of the system's cooling towers was somehow sucked back into the air intake, in most cases due to poor design. This warm air was, needless to say, the perfect environment for the rapid growth of disease-carrying bacteria originating from outside the building, where it existed in harmless quantities...

(The Possible Health Dangers of Air-conditioners[J]. 2002(3):49 - 50.)

上段呈"总-分"结构。

【参考译文】

一个广为人知的危害是莱金奈尔病,20世纪70年代首次得到证实。研究人员发现,空调系统压缩机排出的热气不知怎的(大多由于设计不当)吸入进气口,影响建筑物中人们的健康。不用说,这些温暖的空气为携带疾病的细菌提供了快速增长的完美环境。这些细菌来自外部建筑,以无害的形式大量存在。……

E. g. Although gum cuds can be easily disposed of without creating any problems, gum cuds, when improperly disposed of, can create environmental issues. In this regard the proper disposal of chewing gum, e. g. expectorating the chewing gum on a sidewalk, floor, or like area, can create a nuisance. Due to their typical formulation, gum cuds have an adhesive-like characteristic. Therefore, the chewed gum cuds can stick to environmental surfaces onto which they are intentionally or unintentionally placed...

(傅勇林,唐跃勤.科技翻译[M].北京:外语教学与研究出版社,2012.)

上段呈"总-分"结构。

【参考译文】

虽然轻松地将胶块丢弃看似无虞,但随意乱丢胶块会带来环境问题。对口香糖的不当处理(如将口香糖吐在便道、地板或类似的地方)会产生公害。胶块因其配方独特而具有黏性,因此很容易黏附在一些物体表面上。……

9.3 科技语篇的翻译

9.3.1 科技语篇翻译过程

一般情况下,科技语篇翻译按以下步骤进行:
(1)通读全文,把握段落话题和主要信息点。
(2)标记、翻译生词、短语等。
(3)分析长、难句(分析逻辑关系,划分句子成分)。
(4)逐句翻译,注意以下信息:
1)保证信息的整体性和一致性;
2)做出信息调整(因英汉表意差异);
3)确保译文忠实(于原文)、通顺、流畅。
(5)调整、润色、改进译文。

9.3.2 段落翻译举例

1. 英译汉

Putting the Net to work for the rest of us will be the real challenge in the years ahead. Electronic mail and even videoconferencing are already entrenched, but those applications do not cut to the heart of what the World Wide Web and the rest of the Internet constitute a gigantic storehouse of raw information, the database of all databases. Worries about the future of the Net usually center on the delays and access limitations caused by its overburdened hardware infrastructure. Those may be no more than growing pains, however. The more serious, longer-range obstacle is that much of the information on the

Internet is quirky, transient and chaotically "shelved".

(http://www.doc88.com/p-339760626471.html)

(1)通读全文,把握段落话题和主要信息点。

本段信息呈现"总-分"结构,先总说(主题句):"Putting the Net to work for the rest of us will be the real challenge in the years ahead.",后分述(扩展句),信息不断推进。

(2)标记、翻译生词、短语等。

Putting the Net to work for the rest of us will be the real challenge in the years ahead. Electronic mail and even videoconferencing are already <u>entrenched</u>, but those applications do not <u>cut to the heart of</u> what the World Wide Web and the rest of the Internet constitute a gigantic storehouse of raw information, the database of all databases. Worries about the future of the Net usually center on the delays and access limitations caused by its overburdened hardware infrastructure. Those may be no more than growing pains, however. The more serious, longer-range obstacle is that much of the information on the Internet is <u>quirky</u>, transient and <u>chaotically "shelved"</u>.

相对而言,本段中划线词或短语相对困难,译前需要查词,解决翻译前词汇层困难。

entrenched:根深蒂固的;确立的,不容易改的。(《有道词典》)

一词多义现象存在于几乎所有语言中,而上下文是语义的最终决定因素。我们发现,不宜将《有道词典》提供的"entrenched"直接用来翻译此句,译为"电子邮件甚至视频会议已经根深蒂固的/确立的/不容易改的",因为译文语义不通。可根据上下文,将此句译为:"电子邮件,甚至视频会议已经司空见惯了。"

cut to the heart of:抓住/构成……的实质;直指……的核心。

quirky:古怪的;离奇的。

chaotically "shelved":混乱地"搁置"。在该上下文中可译为"杂乱地'搁置'不用"。

(3)分析长、难句(分析逻辑关系,划分句子成分)。

E.g. Electronic mail and even videoconferencing(主语) are(系动词) already entrenched(表语),but(并列连词) those applications(主语) do not cut(谓语) to the heart(状语) of what the World Wide Web and the rest of the Internet constitute a gigantic storehouse of raw information, the database of all databases(后置定语,呈"of＋宾语从句"结构).

(4)逐句翻译。

(5)调整、润色、改进译文。

E.g. ..., but those applications do not cut to the heart of what the World Wide Web and the rest of the Internet constitute a gigantic storehouse of raw information, the database of all databases.

若照字面翻译,本句意思是"……,但这些应用并不能使互联网和互联网的其他部分成为一个庞大的原始信息和分析数据库,所有数据库的数据库。"译文拗口,且翻译有错误。结合上下文,笔者调整措辞,译为"尽管互联网是所有数据库的数据库,就像一个庞大的原始信息'仓库',但是这些应用并非整个互联网的实质内容。"在保证译文"忠实"和"通顺"的基础上,做到尽量顺译。

E. g. The more serious, longer-range obstacle is that much of the information on the Internet is quirky, transient and chaotically "shelved".

  本句不长,但是不易翻译。《有道词典》将此句翻译为"更严重、更长期的障碍是,互联网上的许多信息都是离奇的、短暂的、混乱的'搁置'"。

  在充分理解原文意思的情况下,笔者调整了信息顺序,将其译为"互联网上的许多信息显得古怪,存在时间短,且杂乱地'搁置'不用,这是互联网发展长期存在的严重障碍。"

【参考译文】

  在未来的岁月里,为他人提供网络服务将成为真正的挑战。电子邮件,甚至视频会议已经司空见惯了,但这些应用并非整个互联网的实质内容,尽管互联网就像一个庞大的原始信息"仓库",是所有数据库的数据库。对网络未来的担忧通常集中在其不堪重负的硬件基础设施所造成的延迟和访问限制上。不过,这些担忧可能只不过是互联网发展中出现的问题而已。互联网上的许多信息显得古怪,存在时间短,且杂乱地"搁置"不用,这是互联网发展长期存在的严重障碍。

2. 汉译英

【例1】

  地震又称地动、地振动,是地壳快速释放能量过程中造成的振动,期间会产生地震波的一种自然现象。地球上板块与板块之间相互挤压碰撞,造成板块边沿及板块内部产生错动和破裂,是引起地震的主要原因。

  地震开始发生的地点称为震源,震源正上方的地面称为震中。破坏性地震的地面振动最烈处称为极震区,极震区往往也就是震中所在的地区。地震常常造成严重人员伤亡,能引起火灾、水灾、有毒气体泄漏、细菌及放射性物质扩散,还可能造成海啸、滑坡、崩塌、地裂缝等次生灾害。

  据统计,地球上每年约发生500多万次地震,即每天要发生上万次的地震。其中绝大多数太小或太远,以至于人们感觉不到;真正能对人类造成严重危害的地震大约有十几二十次;能造成特别严重灾害的地震大约有一两次。人们感觉不到的地震,必须用地震仪才能记录下来;不同类型的地震仪能记录不同强度、不同远近的地震。世界上运转着数以千计的各种地震仪器日夜监测着地震的动向。

  当前的科技水平尚无法预测地震的到来,未来相当长的一段时间内,地震也是无法预测的。所谓成功预测地震的例子,基本都是巧合。对于地震,我们更应该做的是提高建筑抗震等级、做好防御,而不是预测地震。

  (https://baike.baidu.com/item/%E5%9C%B0%E9%9C%87/40588? fr=aladdin)

(1)通读全文,把握段落话题和主要信息点。

  这段文字涉及地震的定义、引起地震的主要原因、地震的构造(震源、震中、极震区)和地震会造成的破坏、地震发生的频率、地震暂时的不可测性等信息。

(2)标记、翻译生词、短语等。

  这段文字相对难以翻译的是地震术语:

地动 earth motion；地壳 earth crust；地震波 earthquake wave；震源 the site where the earthquake occurs；centrum；震中 focus；极震区 meizoseismal area；地裂缝 ground fracture；地震仪 seismometer；抗震等级 seismic grade。

(3)分析长、难句(分析逻辑关系，划分句子成分)。

E. g. 地震又称地动、地振动，是地壳快速释放能量过程中造成的振动，期间会产生地震波的一种自然现象。

翻译界普遍认为，翻译操作应该"自上而下"(即按照篇章—段落—句子—小句—词汇—语音顺序)，高瞻远瞩，能尽可能避免翻译出错。然而，翻译操作往往是"自下而上"(即按照语音—词汇—小句—句子—段落—篇章顺序进行)完成的。

完成"(2)标记、翻译生词、短语等。"后，紧接着就进入译文句式搭建阶段，长句和难句是翻译一大障碍。与英语相比，汉语长句相对少，汉译英的困难主要在译文英语句式的搭建/生成上，即如何生成英语地道的句式，并明示信息之间的逻辑关系。

一般的汉语句子中都含有一个以上的主谓结构。在这个汉语句子中，我们可以划分出"地震又称……""(地震)是……""期间会产生……"等三个主谓结构。翻译时，往往用一个主谓结构作英语句子的主句，让其他主谓结构依照信息之间的逻辑关系设法附着在主句的主谓结构之上。这符合地道英语句子的结构模式。至于用哪个主谓结构作英语译文的主句要视主谓信息之间的重要性来定。英语句式往往呈现"主语＋最重要信息＋废话(其他相对不重要的信息)"结构，所以一般用包含最/更重要信息作英语主句，同时还要考虑其他信息可以附着在英语主句这个主谓之上，还须做到字正句顺，表意清晰。

本句"地震又称……""(地震)是……""期间会产生……"等三个主谓结构中，相对而言"(地震)是地壳快速释放能量过程中造成的振动"包含的信息最重要。"地震又称……"采用顺译法保持原位即可，而"期间会产生……"和其前信息的关系只要明晰即可。因此，本句可译为：An earthquake, also known as earth motion or earth vibration, is the vibration caused by the rapid energy release of earth crust, a natural phenomenon producing an earthquake wave in the process.

E. g. 地震常常造成严重人员伤亡，能引起火灾、水灾、有毒气体泄漏、细菌及放射性物质扩散，还可能造成海啸、滑坡、崩塌、地裂缝等次生灾害。

翻译时，译者一般心怀"能直译要直译""能顺译要顺译"等翻译原则，要知道"笔译是目的语写作"。英语是意合语言，译文中词法、语法、句法须符合英语的表达习惯，这是由英语的本质特征所决定的。

本句较长，是汉语单句，包含一个主语"地震"和三个谓语，呈现"地震造成……，能引起……，还可能造成……"结构。本句含有较多并列成分，顺译即可，所以本句可译为：An earthquake often causes the serious casualties, can cause the fire, floods, toxic gas leak, bacteria and radioactive substance diffusion, and may also cause tsunamis, landslides, collapse of hills, ground fractures and other secondary disasters.

E. g. 当前的科技水平尚无法预测地震的到来，未来相当长的一段时间内，地震也是无法预测的。

本句翻译的困难在于译文句式搭建上。仔细解读"当前的科技水平尚无法预测地震的到来"中的信息关系后我们得知,"预测"的逻辑主语不应是"科技水平",而应是地震工作人员,只是原文未提及(因不是表述的重点内容)。英汉差异告诉我们,这种情况宜将原文汉语转换为英语的被动语态:"Earthquakes cannot be predicted..."。"预测"这个动作的发出者相对内容而言不太重要,原文未提及,译文也尽量不提及,以求译文尽可能贴近原文内容和形式。"地震也是无法预测的"中"无法预测的"对应英语单词"unpredictable",表示性质,与原文表述的意思一致;切不可译为"cannot be predicted",原因有二:①此处表示性质概念而非动作概念;②若这样翻译,将与本句前半部分句子谓语雷同,这是英语一般情况下不允许的。因此,本句可译为:Earthquakes cannot be predicted at the current level of science and technology, and they will be unpredictable for quite a long time in the future.

(4)逐句翻译。

这是语篇翻译过程中非常重要的环节。尽管翻译单位之争方兴未艾,但是大多数学者认同句子这个翻译单位。笔者也认同这一点。句子以下有语音、词汇和小句等信息层,句子以上有段落和篇章层,"进可攻,退可守",句子承载一个完整信息,有利于译者把握意思、解析层次、搭建译文框架并生成译文。

逐句翻译时,译者要心怀该句周围两句的信息,恰当做出判断:将该句译为单句呢?还是与周围句子合译为一句?抑或是将该句拆为两句呢?这样的抉择要"因地制宜",具体情况具体分析,生成准确表意、读来通畅、表达力强、尽可能优美的译文。

(5)调整、润色、改进译文。

这是真正产出高质量译文的阶段。如同好文章一样,好译文都是修改出来。之前四步是为语篇翻译译文生成打下的基础,对初学翻译的人而言必不可少。如何让译文读者对译作的反应尽可能如同原文读者对原作的反应(因措辞准确、逻辑性强、信息推进自然等原因获得清晰的语意甚至美的感受)一样呢?对一般难度的翻译任务而言,大多数错误来自词层。因此,译者在此阶段要反复推敲措辞,力求准确,尤其是要结合上下文用符合目的语表达习惯的词语表意,将问题消灭在萌芽状态。此外,还要检查译文句式的合理搭建,汉译英时还要注意句间关系的明晰情形,保证信息推进自然(往往先总后分,英汉类似),译文文本逻辑性强。

【参考译文】

An earthquake, also known as earth motion or earth vibration, is the vibration caused by the rapid energy release of earth crust, a natural phenomenon producing an earthquake wave in the process. The dislocation and fracture of the plate edge and interior are caused by the extrusion and collision among plates, which is the main reason for the earthquake.

The site where the earthquake occurs first is called the focus, and the ground right above the focus is called the epicenter. The place with the most violent ground vibration in a destructive earthquake is called themeizoseismal area, which is often the area of the epicenter. An earthquake often causes the serious casualties, can cause the fire, floods, toxic gas leak, bacteria and radioactive substance diffusion, and may also cause tsunamis, landslides, collapse of hills, ground fractures and other secondary disasters.

According to statistics, there are more than 5 million earthquakes every year, i.e., over ten thousand earthquakes per day. Most are either too small or too far away, for people to feel. Roughly speaking, there are more than ten or twenty earthquakes per year that cause serious damage and one or two which cause very serious destruction. Earthquakes which people cannot feel must be recorded by seismometers; different types of seismometers can record earthquakes at different intensities and distances. Thousands of kinds of seismic instruments in operation around the world monitor the trend of earthquakes day and night.

Earthquakes cannot be predicted at the current level of science and technology, and they will not be predictable for quite a long time in the future. It is simply coincidental to have the so-called well-predicted earthquakes. Here, what we must do more is improve the seismic grade of buildings with an earnest defense rather than make earthquake prediction.

【例2】

访美期间,习近平主席与奥巴马总统就阿富汗问题深入交换意见,决定就阿富汗问题保持沟通与合作,以支持阿富汗和平重建和经济发展,支持"阿人主导、阿人所有"的和解进程,并促进中、美、阿三边对话。

日前,第四届"美中合作培训阿富汗外交官美方培训班"开班仪式在美国举行。中国外长王毅表示,中美合作培训阿外交官项目增进了中、美、阿三方的相互了解,体现了国际社会支持阿和平重建的共同努力,是中美在第三国合作的成功范例。中美两国还在纽约联合主持,阿富汗重建与区域合作高级别会议,为阿富汗政府和地区经济合作继续提供强有力的国际支持。

(http://news.ifeng.com/a/20151005/44784551_0.shtml)

(1)通读全文,把握段落话题和主要信息点。

这是网上对国家主席习近平2015年9月下旬访美和第四届"美中合作培训阿富汗外交官美方培训班"开班及其意义等的报道。

(2)标记、翻译生词、短语等。

访美期间,习近平主席与奥巴马总统就阿富汗问题<u>深入交换意见</u>,决定就阿富汗问题保持沟通与合作,以支持阿富汗和平重建和经济发展,支持<u>"阿人主导、阿人所有"</u>的<u>和解进程</u>,并促进中美阿三边对话。

日前,第四届"<u>美中合作培训阿富汗外交官美方培训班</u>"开班仪式在美国举行。中国<u>外长</u>王毅表示,中美合作培训阿外交官项目增进了中美阿三方的相互了解,体现了国际社会支持阿<u>和平重建</u>的共同努力,是中美在第三国合作的成功范例。中美两国还在纽约<u>联合主持</u>,阿富汗重建与区域合作<u>高级别会议</u>,为阿富汗政府和地区经济合作继续提供强有力的国际支持。

困难词汇:

深入交换意见 exchange in-depth views;"阿人主导、阿人所有""Afghan-led, Afghan-owned";和解进程 reconciliation process;"美中合作培训阿富汗外交官美方培训班""the US-China sponsored Fourth Afghan Diplomat Training Class";外长 foreign minister;和平

重建 the reestablishment of peace;联合主持 co-chair;高级别会议 high-level meetings。

（3）分析长、难句(分析逻辑关系,划分句子成分)。

E. g. 访美期间,习近平主席与奥巴马总统就阿富汗问题深入交换意见,决定就阿富汗问题保持沟通与合作,以支持阿富汗和平重建和经济发展,支持"阿人主导、阿人所有"的和解进程,并促进中美阿三边对话。

与英语相比,似乎汉语标点使用有时不够科学,也或许是有人汉语写作时标点使用不规范,会出现"一逗到底"现象,即将两个以上信息依然用逗号相连接,尽管应该用句号来连接。在这种情况下,汉译英时要首先以意群为单位合理断句,将汉语原文一个长句拆分为2～3个单句(但不宜再多,否则英语句子信息显得太零散)。

本句应从"……深入交换意见"后拆成两句,这样第一句是总说,第二句为分说,信息有推进。这不仅符合英语信息"先总说,后分说"的布局,而且易于构建英语译文句子结构,信息表述清晰,翻译效果也好。第二句虽然较长,但是英语句子搭建起来不难,"(习近平主席与奥巴马总统)决定……,以支持……,并促进……",采用顺译法即可。因此,本句可译为:During his visit to the United States, Chinese president Xi Jinping exchanged in-depth views on Afghan issues with American President Barack Obama. Regarding those issues, both men decided to maintain communication and cooperation, to promote trilateral dialogue, and to support peaceful reconstruction and economic development in Afghanistan as well as an "Afghan-led, Afghan-owned" reconciliation process.

E. g. 中国外长王毅表示,中美合作培训阿外交官项目增进了中美阿三方的相互了解,体现了国际社会支持阿和平重建的共同努力,是中美在第三国合作的成功范例。

此句较长,但是不难翻译。根据"翻译时尽量保持原文的'风姿'"这一思想,译文英语句子结构可采用原文汉语的结构:"人＋谓语("表示")＋宾语"即可。此句主语是"中国外长王毅",宾语是观点,所以"表示"应译为 said。从汉语原文主要信息是"谁说了什么"来看,"是中美在第三国合作的成功范例"相对不重要(不是主要信息),应该采用弱化手法。因此,本句可译为:China's foreign minister Wang Yi said that, as a successful example of China-US cooperation in third countries, this training program for Afghan diplomats had enhanced mutual understanding among the three sides — China, the United States, and Afghanistan, and that it reflected a joint effort in the reestablishment of peace in Afghanistan with the support of the international community.

（4）逐句翻译。

（5）调整、润色、改进译文。

【参考译文】

During his visit to the United States, Chinese president Xi Jinping exchanged in-depth views on Afghan issues with American President Barack Obama. Regarding those issues, both men decided to maintain communication and cooperation, to promote trilateral dialogue, and to support peaceful reconstruction and economic development in Afghanistan as well as an "Afghan-led, Afghan-owned" reconciliation process.

A few days ago, the opening ceremony for "the US-China sponsored Fourth Afghan

Diplomat Training Class" was held in the United States. China's foreign minister Wang Yi said that, as a successful example of China-US cooperation in third countries, this training program for Afghan diplomats had enhanced mutual understanding among the three sides — China, the United States, and Afghanistan, and that it reflected a joint effort in the reestablishment of peace in Afghanistan with the support of the international community. China and the United States also co-chaired high-level meetings in New York on Afghanistan's reconstruction and on regional cooperation, promising to continue to provide strong international support to the Afghan government and to regional economic cooperation.

练 习 题

一、请将下列文本译为汉语。

Two-speed Internet

Supporters of net neutrality cite two major concerns about these practices. The first is that breaking the internet down into packages renders pricing confusing and difficult to compare, providing cover for mobile operators and ISPs to increase overall costs and pocket the difference.

The second is more systemic: an exclusive list of apps and services that receive preferential treatment divides the internet into haves and have-nots. Sometimes referred to as a "two-speed internet", this runs the risk of entrenching incumbents at the top of the field, while making it very hard for startups to grow to the same scale.

Consider, for instance, trying to advertise a new video-streaming service to New Zealand's Vodafone customers. As well as beating Netflix on its own terms, perhaps by offering better programming or a cheaper subscription, the new service would have to deal with the fact that some Vodafone subscribers will get Netflix without affecting their data caps, but streaming from the startup will eat up their limit very quickly.

But if net neutrality is already weak around the rest of the world, without such negative outcomes becoming widespread, why has the US reacted so strongly — and negatively — to the changes domestically? A glance at Reddit, the self-proclaimed front page of the internet, reveals the scale of the response: 16 of the 25 stories on the site's homepage are about net neutrality, with all but two of them linking to the exact same page.

As San Francisco-based, British-born, venture capitalist Benedict Evans noted, the biggest distinction is competition. "When I lived in London I had a choice of a dozen broadband providers. That made net neutrality a much more theoretical issue," Evans wrote.

In the US, much of the population has essentially no choice over who to buy broadband from, with local monopolies enshrined in law and a nationwide duopoly providing access to high-speed connections for three quarters of the nation. That gives ISPs much more power to wield net neutrality in an extractive fashion, forcing customers to pay extra to access theirfavourite sites at full speed — or forcing companies to pay for access to customers.

The power dynamic has shifted

Barack Obama speaks at a town hall meeting at Google's California headquarters in November 2007. In his visit Obama spoke on his position of net neutrality. Photograph: Kimberly White/Getty Images.

This back and forth has been going on for over a decade, with the term net neutrality coined in 2003, but in recent years the landscape has changed. The most obvious driver has been the election of a US President whose largest motivation for policy decisions appears to be simply undoing the actions of his predecessor.

But the power dynamics between the large internet firms and ISPs has also shifted, in a way that has lessened the institutional support for net neutrality.

In May 2017 Netflix CEO Reed Hastings, once one of the largest proponents of the principle, noted that it didn't really matter for the firm anymore. "Neutrality is really important for the Netflix of 10 years ago, and it's important for society, it's important for innovation, it's important for entrepreneurs… It's not our primary battle at this point…

(https://www.theguardian.com/technology/2017/nov/22/net-neutrality-internet-why-americans-so-worried-about-it-being-scrapped)

二、请将下列文本译为英语。

苹果公司(Apple Inc.)是美国的一家高科技公司。由史蒂夫·乔布斯、斯蒂夫·沃兹尼亚克和罗·韦恩(Ron Wayne)等人于1976年4月1日创立,并命名为美国苹果电脑公司(Apple Computer Inc.),2007年1月9日更名为苹果公司,总部位于加利福尼亚州的库比蒂诺。

苹果公司1980年12月12日公开招股上市,2012年创下6 235亿美元的市值记录,截至2014年6月,苹果公司已经连续三年成为全球市值最大公司。苹果公司在2016年世界500强排行榜中排名第9名。2013年9月30日,在宏盟集团的"全球最佳品牌"报告中,苹果公司超过可口可乐成为世界最有价值品牌。2014年,苹果品牌超越谷歌(Google),成为世界最具价值品牌。

2016年7月20日,《财富》发布了最新的世界500强排行榜,苹果公司名列第9名。

北京时间2016年9月8日凌晨1点,2016苹果秋季新品发布会于在美国旧金山的比尔·格雷厄姆市政礼堂举行。10月,苹果公司成为2016年全球100大最有价值品牌第1名。

2017年1月6日早晨8点整,"红色星期五"促销活动在苹果官网正式上线,瞬间大量用户涌入官网进行抢购,仅两分钟所有参与活动的耳机便被抢光,官网显示"由于需求量大,本次购买的产品不包含 Beats Solo3 Wireless 头戴耳机"。同时苹果官网一度陷入瘫痪，页面无法打开。

　　2017 年 2 月，Brand Finance 发布 2017 年度全球 500 强品牌榜单，苹果公司排名第 2。2017 年 6 月 7 日，2017 年《财富》美国 500 强排行榜发布，苹果公司排名第 3 位。7 月 20 日，2017 年世界 500 强排名第 9 位。

(https：//baike. baidu. com/item/％E8％8B％B9％E6％9E％9C％E5％85％AC％E5％8F％B8/304038？ fr＝aladdin＆. fromid＝6011224＆.fromtitle＝％E8％8B％B9％E6％9E％9C)

下编 实践篇

第 10 章　科普语篇翻译

　　科普,即科学普及(popularization of science),是指以易于大众理解、接受甚至参与的方式向普通民众介绍比较艰难晦涩的自然科学和社会科学相关内容的活动,具体包括推广科学技术的应用、倡导科学方法、传播科学思想、弘扬科学精神等。人类的社会活动与自然、社会科学知识的应用是相辅相成的,两者互为促进,推动社会不断向前发展。在自然科学与人类社会的相互作用中,因自然与人、科学与社会的交叉而产生了科学普及现象和工作,出现了大量承载、传承大众科学知识的读物。

　　因科普读者群、事理深浅度而写作方法等不同,科普语篇读物分为低级、中级和高级三个级别。与科学性强的语篇相比,科普语篇用词较生动,文风较活泼,修辞格使用较多。低级科普读物以启蒙为主,具有激发读者兴趣、启发读者思维、唤起读者求知欲和探索欲等功效,而中、高级科普读物写作目的在于普及科学知识、提供科学方法、推广科学技术等(方梦之,2011)。

10.1　科普语篇的语言特点

(1)在语言风格上,表意形象、生动,文风活泼,修辞格多见。
(2)在语法上,主动语态用得多,句子较短。
(3)在词汇上,多用常见的半技术词,用词形象,富于联想,缩写符号用得少,动词的名词化现象(即用名词短语表达动作概念)少见,动词短语多见。

10.2　科普语篇翻译实例

10.2.1　英译汉

Mars and Water

　　Mars perhaps first caught public fancy in the late 1870s, when Italian astronomer Giovanni Schiapparelli reported using a telescope to observe canali, or channels, on Mars. A possible mistranslation of this word as canals may have fired the imagination of Percival Lowell, an American businessman with an interest in astronomy. Lowell founded an observatory in Arizona, where his observations of the red planet convinced him that the canals were dug by intelligent beings — a view which he energetically promoted for many years.

By the turn of the century, popular songs told of sending messages between Earth and Mars by way of huge signal mirrors. On the dark side, H. G. Wells' 1898 novel *The War of the Worlds* portrayed an invasion of Earth by technologically superior Martians desperate for water. In the early 1900s novelist Edgar Rice Burroughs, known for the *Tarzan* series, also entertained young readers with tales of adventures among the exotic inhabitants of Mars, which he called *Barsoom*.

Fact began to turn against such imaginings when the first robotic spacecraft were sent to Mars in the 1960s. Pictures from the first flyby and orbiter missions showed a desolate world, pockmarked with craters like Earth's Moon. The first wave of Mars exploration culminated in the Viking mission, which sent two orbiters and two landers to the planet in 1975. The landers included experiments that conducted chemical tests in search of life. Most scientists interpreted the results of these tests as negative, deflating hopes of a world where life is widespread.

The science community had many other reasons for being interested in Mars apart from searching for life; the next mission on the drawing boards, Mars Observer, concentrated on a study of the planet's geology and climate. Over the next 20 years, however, new developments in studies on Earth came to change the way that scientists thought about life and Mars.

One was the 1996 announcement by a team from Stanford University, NASA's Johnson Space Center and Quebec's McGill University that a meteorite believed to have originated on Mars contained what might be the fossils of ancient microbes. This rock and other so-called Mars meteorites discovered on several continents on Earth are believed to have been blasted away from the red planet by asteroid or meteor impacts. They are thought to come from Mars because gases trapped in some of the rocks match the composition of Mars' atmosphere. Not all scientists agreed with the conclusions of the team announcing the discovery of fossils, but it reopened the issue of life on Mars.

Other developments that shaped scientists' thinking included new research on how and where life thrives on Earth. The fundamental requirements for life as we know it are liquid water, organic compounds and an energy source for synthesizing complex organic molecules. Beyond these basics, we do not yet understand the environmental and chemical evolution that leads to the origin of life. But in recent years it has become increasingly clear that life can thrive in settings much different from the long held notion of a tropical soup rich in organic nutrients.

In the 1980s and 1990s, biologists found that microbial life has an amazing flexibility for surviving in extreme environments — niches that by turn are extraordinarily hot, or cold, or dry, or under immense pressures — that would be completely inhospitable to humans or complex animals. Some scientists even concluded that life may have begun on

Earth in heat vents far under the ocean's surface.

This in turn had its effect on how scientists thought about Mars. Life might not be so widespread that it would be found at the foot of a lander spacecraft, but it may have thrived billions of years ago in an underground thermal spring. Or it might still exist in some form in niches below the frigid, dry, windswept surface wherever there might be liquid water.

NASA scientists also began to rethink how to look for signs of past or current life on Mars. In this new view, the markers of life may well be so subtle that the range of test equipment required to detect it would be far too complicated to package onto a spacecraft. It made more sense to collect samples of Martian rock, soil and air to bring back to Earth, where they could be subjected to much more extensive laboratory testing with state-of-the-art equipment…

(https://mars.jpl.nasa.gov/msp98/why.html)

这是网络上关于探索火星上是否有水一篇文章的开头内容。本节选文章讲述了从人类对火星充满向往(著书立说、写歌谱曲)到采取实际行动(建天文台观测、将机器人宇宙飞船送上火星、发现并研究陨石等),不断延续着火星生命探索的故事,进一步激发了人类探索火星生命的热情。

翻译这篇节选文章难度不大。本着"能直译要直译,能顺译要顺译"的翻译原则,在准确理解原文信息的基础上,合理构建译文汉语的信息顺序,用地道的汉语流畅地重现原文即可。

这是一篇科普文,原文信息晓畅易懂,文风朴实,叙事性强,几乎没有生僻的专业术语,翻译时要保留原文风格,要注意处理一些易错的细节信息。

1. 词汇层

相对来说,词汇层有以下相对困难的词:

canali(意)渠道;The War of the Worlds《星球大战》;Martian 火星人;Tarzan 泰山(美国影片《人猿泰山》的主人公);pockmark 使有凹坑;microbe 微生物;meteorite 陨石;asteroid【天】小行星;meteor 流星;organic compound 有机化合物;molecule 分子;niche【生态】生态位。

此外,还有几处专有名词:Giovanni Schiapparelli 乔瓦尼·斯基亚帕雷利;Percival Lowell 帕西瓦尔·罗威尔;H. G. Wells H. G. 威尔斯;Edgar Rice Burroughs 埃德加·莱斯·巴勒斯;Quebec's McGill University(加拿大)魁北克省麦吉尔大学。

2. 句子层

句子层的翻译困难往往来自长句和难句。

E.g. A possible mistranslation of this word as canals may have fired the imagination of Percival Lowell, an American businessman with an interest in astronomy.

本句不长,但是翻译时要注意英汉差异转换,否则译文读来不畅。

这是英语单句,句子呈"主语(mistranslation)+谓语(fire)+宾语(imagination)"结构

(括号中为各成分核心词),人名"Percival Lowell"后跟了一个解释性的同位语"businessman",意思理解不难。本句是典型的物称主语句,即主语为物,注重事实,表意客观,而汉语表意时多从人的角度出发描写万物,提倡"天人合一",因此汉语句是典型的人称主语句(很多时候主语为人)。翻译此句时,要注意主语人称和物称的转换,否则读来不流畅。因此,本句可翻译为:"当时,有人可能误将'canali'(意大利语中意为'渠道')译为'canal'('运河'),却激发了美国商人帕西瓦尔·罗威尔对火星的想象力,他对天文学感兴趣。"

E.g. On the dark side, H. G. Wells' 1898 novel The War of the Worlds portrayed an invasion of Earth by technologically superior Martians desperate for water.

本句的翻译翻译困难有二:①"On the dark side"是什么意思?②如何把"technologically superior Martians"译为汉语后,读来更通顺些(而不仅仅是字面的"技术上更优越的火星人")?

"On the dark side"字面意思是"在黑暗的一面",《有道词典》上没有给出这个短语的权威意思。不过,从上下文得知,本句和上一句有些许转折关系。上一句讲述地球和火星之间通过巨大的信号镜传递信息的故事;这句指出小说《星球大战》中火星人入侵地球的故事,具有悲观意义。因此,结合上下文,"On the dark side"可以理解为"We can find the dark side of the stories in between(between Earth and Mars)",然后作者举了小说《星球大战》中的例子。在此,笔者将"On the dark side"翻译为"但是也有人对此并不乐观。"这实际上是将原文英语一个短语译为译文汉语的一句话,发生了翻译单位的变化。这在英汉翻译转换中时有发生,至于翻译单位向上转移(将单词/短语译为句子)还是向下转移(将句子译为短语/单词)要看具体情况来定。

"technologically superior Martians"字面意思是"技术上更优越的火星人",从短语翻译角度讲,这没有问题。笔者一直坚信"笔译是严肃的目的语写作"这个理念,始终认为"句中词/短语服从于句意",在保证译文"忠实"和"通顺"的基础上,在关照译文风格、英汉差异转换妥当的情况下,要尽可能用最高水平的目的语重现原文信息,否则译文质量会受到影响。因此,笔者将本句译为:"但是也有人对此并不乐观。比如,H. G. 威尔斯1898年写的小说《星球大战》讲述了拥有先进科技的火星人渴望获得水源而入侵地球的故事。"这里,"比如"是新增的信息,是为了使译文读来更通畅;"technologically superior Martians"(译为"拥有先进科技的火星人")和"invasion"(译为动词"入侵")翻译时均作了动态化处理(增加动词"拥有"、译作动词),这是因为汉语是动态语言,动词使用频繁。这样处理后,译文字正句顺,丝毫没有"翻译腔",而且很通顺。

E.g. The science community had many other reasons for being interested in Mars apart from searching for life; the next mission on the drawing boards, Mars Observer, concentrated on a study of the planet's geology and climate.

翻译本句时要注意细节。"The science community"不可译为字面的"科学共同体",这与上下文不符,而应译为"科学界";"the next mission on the drawing boards"中"on the drawing boards"为比喻说法,意为"处于设计阶段的",而非"在画板上的"。所以笔者将本句翻译为:"除了寻找火星生命,科学界对火星感兴趣还有很多其他原因;处于设计阶段的下一项任务由'火星观察者号'探测器执行,主要侧重于研究火星的地质和气候。"

E. g. This rock and other so-called Mars meteorites discovered on several continents on Earth are believed to have been blasted away from the red planet by asteroid or meteor impacts.

　　翻译本句时要注意原文英语句谓语部分被动语态的转化工作。

　　汉语表意时很少用"被"字句,除非为了强调效果。比如,"我被冤枉了。"这个说法可能是为了获取别人同情。英译汉时,要尽可能地去除"被"字眼,因为中、高级英语为表意客观多用被动语态,而汉语多用主动语态表意。根据上下文,"believe"的动作发出者是上一句提到的"斯坦福大学、美国航空和航天局约翰逊航天中心和加拿大魁北克省麦吉尔大学组成的一个研究团队",所以笔者在本句添加"believe"的动作发出者"他们",作汉语主语,这样就将原文被动句译为汉语主动句:"他们认为,这块石头和其他几块在地球大陆上发现的所谓的火星陨石,可能因小行星或流星撞击火星产生爆裂而掉落在地球上。"

E. g. In the 1980s and 1990s, biologists found that microbial life has an amazing flexibility for surviving in extreme environments — niches that by turn are extraordinarily hot, or cold, or dry, or under immense pressures — that would be completely inhospitable to humans or complex animals.

　　本句为英语长句,为主从复杂句,由主句"biologists found that microbial life has an amazing flexibility for surviving in extreme environments"连同其后的两个"that"从句构成。其中,"niches"为"environments"的同位语,是其后两个"that"从句的先行词。

　　英语为形合语言,信息之间关系须明示,各信息亦可有自己的修饰成分,句子似乎可以无限延长;汉语一般用短句表意,句子随表随释,主谓结构并列现象多。为译文读来流畅起见,笔者将本句翻译为:"20 世纪 80 年代和 90 年代,生物学家发现微生物具有极强的适应性,可以在极端恶劣的环境(交替出现极热、极冷、极干燥或气压极强的生态情况)中生存,而人类或复杂动物不宜在这种环境下生活。"其中,笔者将第一个"that"从句弱化为解释性信息,译入括号中;笔者将第二个"that"从句译为一个主谓结构句,与原文主句并列,语义上形成鲜明的对比。

E. g. It made more sense to collect samples of Martian rock, soil and air to bring back to Earth, where they could be subjected to much more extensive laboratory testing with state-of-the-art equipment...

　　翻译本句时,同样要注意英汉差异的恰当转换。英语表态在先(如本句"It made more sense to..."),叙事在后;汉语恰恰相反。因此,笔者将本句翻译为:"……而先收集火星岩石、土壤和大气样本并带回地球,然后在实验室使用最先进的设备对其开展更广泛的测试显得更加可行一些。"这样处理后,译文表述即符合汉语的表达习惯了。

【参考译文】

火星和水

　　19 世纪 70 年代后期,人类可能第一次开始对火星充满向往。当时,意大利天文学家乔瓦尼·斯基亚帕雷利(Giovanni Schiapparelli)对外宣称他使用望远镜观察到了火星上的渠道,或者说水道。当时,有人可能误将"canali"(在意大利语中意为"渠道")译为"canal"("运河"),却激发了美国商人帕西瓦尔·罗威尔（Percival Lowell）对火星的想象力,他对天文学感兴趣。之后,罗威尔在美国亚利桑那州建造了一个天文台。在那里,他对这颗红色星球进行观察,最终认定火星上的运河是外星人挖掘的。此后很多年,他积极宣扬这一观点。

到了19世纪末20世纪初,很多流行歌曲讲述在地球和火星之间通过巨大的信号镜传递信息的故事,但是也有人对此并不乐观。比如,H. G. 威尔斯1898年写的小说《星球大战》讲述了拥有先进科技的火星人渴望获得水源而入侵地球的故事。20世纪早期,因《泰山》系列小说而知名的小说家埃德加·莱斯·巴勒斯（Edgar Rice Burroughs）也创作了火星居民的冒险故事《巴松》,受到年轻读者的追捧。

20世纪60年代,人类第一次将机器人宇宙飞船成功送上火星,事实开始打破人类的无尽想象。从飞船执行第一次飞近火星探测和轨道飞行任务发回的照片来看,火星上非常荒凉,表面布满了坑,跟地球的卫星月球很像。1975年,在执行维京任务（向火星发送了两个轨道飞行器和两个着陆器）期间,第一波火星探险达到高潮。着陆器还要执行一项任务,那就是在火星上进行化学实验,探测是否有生命存在。大部分科学家认为这些实结果证明火星上没有生命,这对于持"火星上广泛分布着智慧生命"观点的人们来说是一个打击。

除了寻找火星生命,科学界对火星感兴趣还有很多其他原因;处于设计阶段的下一项任务由"火星观察者号"探测器执行,主要侧重于研究火星的地质和气候。但是,在之后的20年间,地球研究的新发现改变了科学家以往对生命和火星的认知。

1996年,斯坦福大学、美国航空和航天局约翰逊航天中心和加拿大魁北克省麦吉尔大学组成的一个研究团队宣布他们发现了一块可能来自火星的陨石,可能含有古代微生物的化石。他们认为,这块石头和其他几块在地球大陆上发现的所谓的火星陨石,可能因小行星或流星撞击火星产生爆裂而掉落在地球上。科学家认定这些陨石来自火星,因为有几块石头内含有的气体与火星大气层的成分相同。虽然并不是所有的科学家都同意这个团队得出的结论（发现化石）,不过由此重新引起了人们对火星生命的讨论。

对科学家认知产生影响的其他新进展还包括针对地球充满生机的方式和地点展开的新研究。众所周知,生命发展的最基本需求是水、有机化合物以及用于合成有机分子的能量来源。除了这些基本要素,我们还不明白形成生命起源的环境和化学进化情况。但是近年来,人们越来越清楚地认识到,除了我们长期以来认为的富含有机物的热带环境,生命还可以在与之截然相反的环境中生存并蓬勃发展。

20世纪80年代和90年代,生物学家发现微生物具有极强的适应性,可以在极端恶劣的环境（交替出现极热、极冷、极干燥或气压极强的生态情况）中生存,而人类或复杂动物不宜在这种环境下生活。有些科学家甚至得出这样的结论:地球上的生命可能发源于距离海平面以下极深处的火山口。

这反过来又影响了科学家对火星的看法。火星上的生命不可能分布得如此广泛,以至于飞船一登上火星就会发现生命体,但是几十亿年前,在火星上某一处地下温泉里生命也许蓬勃发展过。或者,在寒冷、干燥、风吹的地表下有液态水的地方,生命以某种方式仍然存在着。

美国航空和航天局的科学家也开始重新思考寻找火星上存在过或存在着生命痕迹的方法。新的观点认为,生命标志可能太过于微小以至于用来检测它的检验设备太过于复杂而不能装进一艘宇宙飞船,而先收集火星岩石、土壤和大气样本并带回地球,然后在实验室使用最先进的设备对其开展更广泛的测试显得更加可行一些。

10.2.2 汉译英

药用植物闭鞘姜

文章来源:西双版纳热带植物园;发布时间:2014-10-30

闭鞘姜〔Costus speciosus(Koen.)Smith〕为闭鞘姜科闭鞘姜属多年生草本,别名广商陆、水蕉花、老妈妈拐棍;株高 1~3 米,基部近木质,顶部常分枝,旋卷;叶片长圆形或披针形,长 15~20 厘米,宽 6~10 厘米,顶端渐尖或尾状渐尖,基部近圆形,叶背密被绢毛;穗状花序顶生,椭圆形或卵形,长 5~15 厘米;花期 7—9 月,果期 9—11 月。

闭鞘姜产于我国广东、广西、云南等省区;生于疏林下、山谷荫湿地、路边草丛、荒坡、水沟边等处,海拔 45~1 700 米;热带亚洲广布。

闭鞘姜根茎有小毒,但可供药用,有消炎利尿、散瘀消肿的功效。

闭鞘姜是西双版纳的一种土著植物,当地多种少数民族都对它比较熟悉,且均以之入药,比如:傣族用它治疗热风咽喉肿痛、腮腺和颌下淋巴结肿痛、化脓性中耳炎、冷风湿关节疼痛所致曲伸不利、肢体关节红肿疼痛等病症;哈尼族用它治疗中耳炎、肾炎水肿等病症;基诺族用它治疗结石、泌尿系统感染、水肿、荨麻疹、疮疖、肿毒等病症。

(http://www.cas.cn/kxcb/kpwz/201410/t20141030_4233260.shtml)

本文讲述了药用植物闭鞘姜的结构、产地及其药物功效等信息。

从本文可见汉语典型的表述方式:用短句表意,很少用关联词,信息间逻辑靠内部关系关联,信息呈竹节状,不断延伸。

1. 词汇层

词层的翻译困难主要来自跟植物有关的一些术语和医药类一些术语:

闭鞘姜 Costus speciosus(Koen.)Smith;广商陆 Guangshanglu;木质 xylogen;披针形 lanceolate;绢毛 fine silky hair;穗状花序顶生 spike terminal;椭圆形 ellipse;卵形 oval;果期 fruiting season;消炎利尿、散瘀消肿 anti-inflammatory for treating diuresis and stasis as well as for reducing swelling;热风咽喉肿痛 sore throat caused by wind and heat;腮腺肿痛 sore parotid;颌下淋巴结肿痛 sub-mandibular lymph glands;化脓性中耳炎 purulent middle-ear inflammation;关节疼痛所致曲伸不利 stiffness and joint pain due to rheumatoid arthritis;肢体关节红肿疼痛 inflamed swollen joints。

2. 句子层

句子层的翻译困难来自译文英语句式的合理搭建和信息之间关系的明晰。

E. g. 株高 1~3 米,基部近木质,顶部常分枝,旋卷。

如何明晰三个逗号前后的关系呢?

本句包含三个主谓结构,即"株高 1~3 米""基部近木质"和"顶部常分枝,旋卷",意思相对独立,这是汉语典型的表意手法,但是英语为形合语言,将本句译为英语时不可译为三个英语单句,否则译文可读性极低。可将第一个逗号前后信息视作并列关系,第二个逗号前后关系视作比较,用"while"连接,这样符合明示逻辑关系(即所谓"形合")英语的表达习惯,读来也朗朗上口。因此,本句可译为:"It is normally 1 - 3 meters tall and its base consists almost entirely of xylogen, while its top usually branches spirally."

E. g. 叶片长圆形或披针形,长 15~20 厘米,宽 6~10 厘米,顶端渐尖或尾状渐尖,基部近圆形,叶背密被绢毛。

同样,本句中还有多个主谓结构,然而汉语却一逗到底(总共用了 5 个逗号),切不可断为多个英语句子,否则可读性极低。因此,翻译本句的困难同样来自英文句式的搭建和逻辑关系的明晰上。

笔译是严肃的目的语写作。中、高级英文写作时,要尽可能将相关信息写在一句中,要主次分明,这样易于读者获取信息。但是,当无法明晰关系,即要断成另一句,不可"因噎废食"(为写长句而写长句),写出错误句。对于汉语逗号多、稍长点句子,译者要以意群为单位断句,搭建合理英语句型,析出并明示逻辑关系,即可大功告成。

本句中的"叶片长圆形或披针形,长 15~20 厘米,宽 6~10 厘米,顶端渐尖或尾状渐尖……"描述了这种植物的叶片形状、长度和宽度变化情形,包含 3 个意思。通过挖掘信息间的逻辑关系,笔者发现可将这 3 个意思写为有层次感的一句英语:"Its oblong or lanceolate leaves, which are 15 - 20 cm long and 6 - 10cm wide become gradually sharper at the head and tail." "基部近圆形,叶背密被绢毛"陈述了"基部"和"叶背"两个概念,可分两句译出,第二句"叶背"信息与下边信息连为一句,将英语句子拉长。本句可译为:"The base part is nearly round. The backs of the leaves are thickly covered with fine silky hairs..."

E. g. 生于疏林下、山谷荫湿地、路边草丛、荒坡、水沟边等处,海拔 45~1 700 米;热带亚洲广布。

翻译本句时要注意"疏林""山谷荫湿地""路边草丛""荒坡""水沟边"这些概念英语准确的表述方法,而且要合理安排信息顺序和明晰逻辑关系。本句可译为:"Being widespread in tropical regions in Asia, it grows mainly in open forests, shady valleys, wetlands, roadside underbrush, on desolate slopes, and on the edges of ditches at altitudes ranging from 45 to 1,700 meters." 从中可以看出信息的层次和信息呈现的多样性。

E. g. 闭鞘姜根茎有小毒,但可供药用,有消炎利尿、散瘀消肿的功效。

本句中"但"表明其前后信息呈转折关系,"可供药用"和"有消炎利尿、散瘀消肿的功效"是因果关系,前果后因,所以本句可译为:"The rhizomes of Costus speciosus (Koen.) Smith are slightly toxic but can be used for medical purposes, since they are therapeutically effective as an anti-inflammatory drug for treating diuresis and stasis as well as for reducing swelling."

E. g. 傣族用它治疗热风咽喉肿痛、腮腺和颌下淋巴结肿痛、化脓性中耳炎、冷风湿关节疼痛所致曲伸不利、肢体关节红肿疼痛等病症;哈尼族用它治疗中耳炎、肾炎水肿等病症;基诺族用它治疗结石、泌尿系统感染、水肿、荨麻疹、疮疖、肿毒等病症。

本句包含很多医药术语,并列呈现,英文可照猫画虎,一一罗列即可。在这里,原文汉语有重复表述的字眼,如"族""治疗",英语表述要避免重复,所以翻译时要尽可能用不同表述法。

"傣族"可译为"the Dai ethnic group"(此处指人),"哈尼族"可译为"The Hani people","治疗"可用"treat"和"cure"交替表述(本句译文见下文)。

【参考译文】

The Medical Plant — Costus speciosus(Koen.)Smith

Source:Xishuangbanna Tropical Botanical Garden   Date:October 30, 2014

Costus speciosus(Koen.)Smith, also known as Guangshanglu, Water Banana Flower or Granny's Walking Stick, is a perennial herbaceous plant of the Costus family. Normally 1 – 3 meters tall, its base consists almost entirely of xylogen, while its top usually branches spirally. Its oblong or lanceolate leaves, which are 15 – 20 cm long and 6 – 10 cm wide become gradually sharper at the head and tail. The base part is nearly round. The backs of the leaves are thickly covered with fine silky hairs and the 5 – 15 cm long spike terminals of its flower clusters take the shape of an ellipse or oval. This plant comes into bloom from July to September and its fruiting season lasts from September to November.

In China, Costus speciosus (Koen.) Smith grows chiefly in Guangdong Province, Guangxi Province and Yunan Province, etc. Being widespread in tropical regions in Asia, it grows mainly in open forests, shady valleys, wetlands, roadside underbrush, on desolate slopes, and on the edges of ditches at altitudes ranging from 45 to 1,700 meters.

The rhizomes of Costus speciosus (Koen.) Smith are slightly toxic but can be used for medical purposes, since they are therapeutically effective as an anti-inflammatory drug for treating diuresis and stasis as well as for reducing swelling.

This plant is indigenous to Xishuangbanna in China, so most of the local minorities are very familiar with it and even use it as a medicine. For example, the Dai ethnic group use it to treat conditions such as sore throat caused by wind and heat, sore parotid and swelling of sub-mandibular lymph glands, purulent middle-ear inflammation, stiffness and joint pain due to rheumatoid arthritis, inflamed swollen joints, etc. The Hani people use it to cure diseases like otitis media and nephritic edema, while the Jinuo people use it to treat such diseases as lithiasis, urinary tract infection, edema, urticaria, furunculosis, and toxic swelling.

练 习 题

一、请将以下文本翻译为汉语。

TV Addiction — A Growing Problem

Most people of my generation have never known life without television. We have grown up sitting in the living room in front of the flickering TV screen. It is hard for us to see how television affects our lives. The belief that television is destructive to communication among family and friends is quite common. On the other hand, television has increased the speed at which information travels, and it can transfer circumstances and images into our homes that previous generations never had. People tend to idealize the past; they imagine a group of family and friends entertaining themselves by playing games and telling stories around a warm fireplace, but I do not think that TV can be blamed for the lack of communication among family and friends.

Without a doubt, television is one of the most powerful means of communicating in the past decade, competing with other forms of communication such as the Internet, telephone, movies, cellular phones, and of course, our speech. Due to its extensive availability and enriched media with images and sounds, it is very difficult to keep it away from our lives. Television is a necessity of our lives like our meals, clothing, and home.

At first, the broadcasting industry was started for public purposes. Nowadays, as to the influences of television, some people say that television offers a relaxing time and useful information to us while others argue that television does not always have a good effect on us because of its hindrance to communication among people. However, television provides us with rich topics to talk about and a chance to get together with each other. Most of all, television helps us communicate with others. We watch television almost every evening with our family. For example, my father and I enjoy watching news or knowledge-based documentary programs. Watching these programs we sometimes discuss current affairs such as the Presidential election or economic problems. Likewise, hot issues on television like news or trendy programs are the most common topics among my friends in my college. If some of them did not watch television the previous night, they cannot even take part in the discussion because they have nothing to talk about. In this respect, since television provides us with various topics related to our lives, it helps us form our sense of intimacy with family or friends.

Next, television promotes our communication because it offers a chance to get together with families or friends. Nowadays, most people are so busy that they do not have much time to spend with family and friends. However, through TV programs, we can have a chance to get along with them. For instance, when the 2002 World Cup was held in Korea, many people had a great time with their family or friends. To watch soccer games, we gathered in front of TV with friends or family who we were alienated from for a long time. Consequently, such TV programs help us build up closer human relationships among people. Thus, I think television plays an important role in promoting communication with family or friends.

"Children are more likely to become actively engaged with anything that attracts them. Especially, television provides great interests with its variety of sounds and images to children. They immediately desire for watching it whether content is neither too easy not too difficult to understand. By its providing some challenge, television allows them to gain an abundant vocabulary. Just like our muscles, the brain gets stronger when it is used, and declines when it isn't used. Television is so commonly criticized as being bad for children that an important fact sometimes gets overlooked". (Science Daily, 2001)

Some types of television viewing may actually enhance children's intellectual development. Television contains an enormous variety of forms and content. In a study, the effects of television viewing depended on program content.

There are always bad effects on children watching TV. Children who are TV viewers increasingly view life as an entertainment extravaganza in which they are fond of playing role in TV show, and there are aggressive content of current broadcasting that greatly affect children. Besides, it is inevitable to watch TV without facing revolting violence. Once more, the effects of television viewing depend on program content and genre. Children are required to be advised what right TV program to watch by their parents. (Carter Bill, 1996)

(https://www.ukessays.com/essays/general-studies/tv-addiction-problem.php)

二、请将以下文本翻译为英语。

城市绿地是城市生物多样性的重要载体，是反映一个城市生活质量的重要指标，对保障城市生态环境的可持续发展和维护居民身心健康起着至关重要的作用。土壤是绿地系统的一个重要组成要素，既是植被的立地基础和生长介质又是城市生态系统中物质转化与能量循环的必要环节。

近年来，一些学者遴选了表征土壤质量的指标，运用高层次综合、多指标量化手段，建立了土壤质量综合指标体系和评价系统，并对土壤在时间和空间上的演变规律进行了研究，但这些研究更多关注的是农业耕作土壤，对于城市绿地土壤的物理性能评价较少。

城市新建绿地土壤不同于农业耕作土壤与森林草原土壤，它受到人类活动影响的深度与频度更为剧烈，这些活动最终累积在土壤物理结构的时空变异上，形成土壤质量的隐形退化。土壤物理退化制约植物根系的延伸空间和延伸速率，决定着植物对于不良环境的抵御能力，表现在城市绿地生态系统的功能紊乱，景观效应的单调与雷同，生物多样性退化等方面。

土壤质量是土壤内部功能与外部环境之间进行相互作用的能力，土壤协调营养和环境的能力主要依赖于土壤胶体和土壤团聚作用，即土壤的物理状态。对城市绿地土壤质量的评价如果仅着眼于土壤肥力,就无法突出评价目标的特殊性,评价结果也是片面性的。Karlen等认为,土壤质量评价具有时空差异性和环境特殊性,土壤质量评价指标体系的确立需要明确问题发生的区域。不同地区、不同景观类型的土地也应使用不同的指标体系,要解决指标体系、评价方法与研究尺度三者的适宜性。

笔者以西安植物园新园区为典型案例,对园区内土壤的物理性质进行调查研究,拟从土壤物理性质的空间变异,结合模糊数学方法对城市新建绿地土壤的物理质量进行评价,阐述新建绿地土壤各层间的差别,分析各因子的现状及相互关系,以期为系统评价西安市城市新建绿地土壤的物理特性提供必要的参考,进而为西安城市生态环境的可持续发展、海绵城市的建设提供基础数据。

(丛晓峰,余刚,谢斌,等. 基于模糊数学法的新建绿地土壤物理性能评价[J]. 中国农学通报,2018(29):59-63.)

第 11 章　学术论文摘要翻译

11.1　学术论文摘要的构成及其特点

11.1.1　摘要的构成

摘要,顾名思义,就是"摘录的要点"的意思,是学术论文必不可少的一部分。在现今论文在线发表的社会语境下,摘要的写作和发表可方便学者在线检索相关论文,为研究者提供研究思路、拓宽研究视域甚至解决相关问题提供帮助,提高学术研究的效率,具有重要意义。

在学术界,摘要的写作有严格的规定。一般来说,摘要须包含以下四个部分。

(1)目的。即研究者做这项研究的目的或要解决的问题,这是一篇好文章在"引言"部分需要首先明确交代的问题。一般情况下,提出研究目的前需要简介前人的最新研究成果,在此基础上,"你"想解决什么问题。

(2)方法。在此主要介绍为达到研究"目的","你"所使用的方法及边界条件(如使用的主要设备和仪器等)。社会科学研究方法一般包括定性方法、定量方法、文献调查法、抽样方法、访谈法、问卷调查法、观察法、思辨法、行为研究法、历史研究法、概念分析法、比较研究法等;自然科学研究方法一般包括科学实验法、数学方法、信息方法、系统科学方法、控制论方法以及一些复杂科学研究方法(如耗散结构理论、协同学、混沌理论等)。

(3)结果。这里陈述研究得出的结果和发现。

(4)结论。也就是通过研究得出的结论,与"结果"一脉相承,有直接关联性,但在思想上高于"结果",是"结果"的升华。这是论文写作非常重要的内容,因为这是研究工作/论文对相关领域的主要成就和贡献部分,直接关乎该研究/论文的价值,是读者最想了解的信息。

11.1.2　摘要的特点

摘要应具备以下特点:

(1)完整性。研究论文的摘要应完整包含研究的目的、方法、结果和结论,这四部分浑然一体,互为支撑,缺一不可。

(2)客观性。摘要须与论文正文完全一致,不可夸大,也不可缩小。

(3)高度的概括性。摘要是论文的核心内容,浓缩的是论文的"精华"。

(4)简洁性。摘要表述要简洁,尽可能用简单明了的话语准确地表述有关概念,表达丰富的研究内容,不啰唆。

11.2 学术论文摘要翻译实例

11.2.1 英译汉

Exploring the Author-Translator Dynamic in Translation Workshops

×××

Abstract

The traditional notion of the translator as someone who should remain invisible while reproducing the original and/or intentions of the author is still commonplace today in translation workshops. Although it has been radically called into question by poststructuralist theory, this type of theory often does not "translate" into what students understand as the practice of the craft. The essay draws on a comparative study used with the author's students that involved eight English versions of Jorge Luis Borges's 1960 text "*Borges y yo*" to indirectly introduce them to poststructuralist notions of translation, reading and authorship that can help them confront the limitations of the traditional conception of translation and assist them in developing the critical capacity to work responsibly through the complexities involved in the task of rewriting someone else's text in another language. This activity-with its combination of close readings of the eight translations together with an analysis of the text's plot in the context of the contemporary notion of the "death of the author" — helps students discover that they cannot escape complex ethical decisions related to their agency both as readers of an "original" and as authors of their translations, even when, as is the case with one of the translations, the author has collaborated with the translator.

(WYKEB V. Exploring the Author-Translator Dynamic in Translation Workshops[J]. The Translator, 2012, 18(1): 77 – 100.)

摘要是严肃的文本写作类型，一般就四部分内容（即目的、方法、结果和结论）简洁、清晰描述研究工作，没有模棱两可的话语，所以理解起来困难不大。

1. 词汇层

dynamic 动态；poststructuralist theory 后结构主义理论；call into question 对……表示怀疑；Jorge Luis Borges 乔治·路易斯·博尔赫斯；*Borges y yo*《博尔赫斯和我》；collaborate 合作。

2. 句子层

E.g. The traditional notion of the translator as someone who should remain invisible while reproducing the original and/or intentions of the author is still commonplace today in translation workshops.

本句为"主语＋系动词 be＋表语"结构，主语较长。本句翻译理解主语是关键。因为英语是形合语言，所以逻辑关系明示，很方便读者阅读理解。本句主语核心词是"notion"，其后

"of"短语起修饰、限定该核心词作用,"as"意为"作为","who"引导定语从句,修饰"someone"。理解了这些明示的逻辑关系后,接下来就是按照汉语的表达习惯做连词成句工作了。笔者将本句翻译为:"在重现原文及/或作者意图的过程中,译者应该保持隐形这一传统观念在当今翻译工作坊中依然普遍存在。"

E. g. The essay draws on a comparative study used with the author's students that involved eight English versions of Jorge Luis Borges's 1960 text "*Borges y yo*" to indirectly introduce them to poststructuralist notions of translation, reading and authorship that can help them confront the limitations of the traditional conception of translation and assist them in developing the critical capacity to work responsibly through the complexities involved in the task of rewriting someone else's text in another language.

本句是这篇摘要中最长的句子。英译汉的困难往往在阅读理解上,译者要有足够的耐心去理解信息之间的逻辑关系,将信息顺序转换为汉语的信息顺序。不过,由于英语信息之间的逻辑关系都是外显(明示)的,所以原文英语句子即使较长,只要译者有足够耐心,理解原文的困难也不会太大。

就拿这个句子来说,划分句子成分后,我们就知道本句的主干信息是"The essay draws on a comparative study"〔意思是"本文运用比较研究(法)……"〕,"to indirectly introduce them poststructuralist notions of translation"(意思是"间接向学生介绍后结构主义的翻译观点")表明开展这项研究的目的,其后的"reading and authorship..."是"后结构主义的翻译观点"的并列成分。翻译界同仁认为,当构成定语从句的字眼大于8个,译为汉语时要断为汉语当句。"reading and authorship"后"that"引导的定语从句构成字眼远大于8个,所以与前一句断开,译为另一句即可。笔者将本句翻译为:"本文比较了笔者学生八个版本《Borges y yo》(《博尔赫斯和我》)〔乔治·路易斯·博尔赫斯(Jorge Luis Borges),1960年出版〕的翻译作品,间接向学生介绍后结构主义翻译观点、阅读及了解作者背景信息的重要性,这不仅有助于译者突破传统翻译理念的局限性,而且在译者用另一种语言翻译他人文本过程中,有助于译者获得处理复杂问题所需的重要能力。"

【参考译文】

翻译工作坊中作者-译者之间的动态关系探究
×××

摘要

在重现原文及/或作者意图的过程中,译者应该保持隐形这一传统观念在当今翻译工作坊中依然普遍存在。尽管后结构主义理论对这一现象提出质疑,但是学生在实践中却无法把理论"转化"为自身的理解。本文比较了笔者学生八个版本《Borges y yo》(《博尔赫斯和我》)〔乔治·路易斯·博尔赫斯(Jorge Luis Borges),1960年出版〕的翻译作品,间接向学生介绍后结构主义翻译观点、阅读及了解作者背景信息的重要性,这不仅有助于译者突破传统翻译理念的局限性,而且在译者用另一种语言翻译他人文本过程中,有助于译者获得处理复杂问题所需的重要能力。通过上述活动以及对八个翻译版本的比较,笔者借助当代"作者之死"概念分析文本情节内容,帮助学生明白他们作为"原著"读者和译著作者这种中介身份无法避免做出复杂的道德抉择,即便作者与译者合作翻译,情况也是如此。

11.2.2 汉译英

【例1】

民主转型与政治暴力冲突的起落：以印尼为例

×××

摘要

相较于"稳固时期的威权统治"，民主转型初期常常是一个政治暴力风险上升的历史阶段。本文指出，尽管民主化同时蕴含暴力加剧和暴力缓解的逻辑机制，但在一个族群分裂的社会，由于民主化初期政治信任格外匮乏，其暴力加剧机制常常早于暴力缓解机制出现。那么，为什么即使在社会分裂的国家，转型期不同国家的暴力冲突水平仍然出现明显差异？即，为什么一些国家比另一些国家暴力冲突水平高得多？本研究聚焦于政治宽容这个因素，以此来解释多族群社会转型暴力水平的差异。在本文中，政治宽容包括大众文化和精英意识两个维度，并强调其相辅相成性。简单而言，政治宽容水平高的地方，暴力冲突水平低，反之则否。本文以印尼为例，对核心观点进行了论证与说明。

（刘瑜. 民主转型与政治暴力冲突的起落：以印尼为例[J]. 学海，2017(2)：45－55.）

1. 词汇层

词汇层的翻译困难往往是词汇量小和词汇没有恰当使用的问题，看似简单，但可能是翻译错误最大的来源。词汇的正确使用与否彰显译者的基本功。

这篇摘要相对困难的词汇如下：

民主转型 democratic transition；政治暴力 political violence；起落 ups and downs；威权统治 authoritarian rule；暴力加剧 violence intensification；暴力缓解 violence alleviation；族群分裂 ethnic division；政治宽容 political tolerance；大众文化 mass culture；精英意识 awareness of elites；维度 dimension。

2. 标题

一般情况下，标题翻译要尽量避免句子形态（即译为句子），而要译为短语形态，因为短语可突出核心信息，句子次之。

本标题"民主转型与政治暴力冲突的起落：以印尼为例"中最重要信息"起落"不言自明，关键是要从其前的"民主转型与政治暴力冲突"理解出相对重要的信息是"冲突"，发生在"民主转型"和"政治暴力"之间。本标题的核心概念是"冲突的起落"，至此，逻辑关系就明晰了，翻译操作也变得简单了。本句可译为："Ups and Downs of Conflicts between Democratic Transition and Political Violence：Taking Indonesia as an Example"。

3. 句子层

译者翻译句子时，要心怀周围两句的存在（有时需合译，汉译英时常常需要明晰句间关系），合理搭建译文框架/句型，恰当转换英汉差异，用自己能驾驭对的最高水平写作能力重现原文信息。

E.g. 相较于"稳固时期的威权统治"，民主转型初期常常是一个政治暴力风险上升的历史阶段。

汉语句子常常呈现"主语（有时缺失）＋废话（相对不重要信息，如状语）＋谓语（最重要信息)"结构,而英语句子常常呈现"主语(有时缺失)＋谓语(最重要信息)＋废话(相对不重要信息,如状语)"结构。由此可以看出,汉语句子重要信息靠近句末(先将信息铺垫做充分),而英语句子重要信息靠近句首。因此,汉译英时,为快速、准确搭建英语句子结构,可将原文一般情况下不长的汉语句子缩到最短,作为译文英语的主句,让其他信息设法附着在这个做主句的主谓结构之上。

本句不长,主句缩至最短即为"初期是阶段",其他信息各就各位即可。

另外还要注意,相对于英语而言,汉语表意的模糊性时有存在。因此,汉译英时,需将汉语的模糊概念转化为清晰的英语概念。本句中的"民主转型"须译作"a transition to democracy";在"一个政治暴力风险上升的历史阶段"中,"政治暴力风险上升"和"历史阶段"尽管仅仅用"的"连接,但是细读起来,我们会发现它们之间的关系并非只是从属关系,而是前者是后者的特点,所以最清晰明示两者之间关系的做法是用"characterized by"连接。本句可译为:"Compared with 'authoritarian rule in a steady period', the initial period of a transition to democracy is a historical stage characterized by increasing risk of political violence."

E.g. 本文指出,尽管民主化同时蕴含暴力加剧和暴力缓解的逻辑机制,但在一个族群分裂的社会,由于民主化初期政治信任格外匮乏,其暴力加剧机制常常早于暴力缓解机制出现。

理想的翻译如同照相,这样原文所有的信息就原封不动地保留到了译文里。翻译时,尽可能照猫画虎,尽量依照原文信息出现的顺序一一罗列,能直译则直译,不能直译则变通处理。基于此,本句可译为:"This paper points out that although democratization contains a logical mechanism of violence intensification and violence alleviation, political trust in the early period of democratization is particularly hard to come by in a society with ethnic divisions and therefore, the mechanism of violence intensification often appears earlier than the mechanism of violence alleviation." 其中,将在原文中作原因状语的"由于民主化初期政治信任格外匮乏"处理为译文宾语从句的主句的前半部分,与其后的"其暴力加剧机制常常早于暴力缓解机制出现"在译文中并列呈现,增加关联词"therefore",这样逻辑清晰,说理透彻。

E.g. 在本文中,政治宽容包括大众文化和精英意识两个维度,并强调其相辅相成性。

原文中的"包括"和"强调"若分别直译为"contain"和"emphasize",将导致英语译文读来生硬,表意不自然。一般情况下,英语表意时往往措辞委婉,读来语气舒缓。本句可译为:"In this paper, political tolerance is understood as having two complementary dimensions — 'mass culture' and 'awareness of elites', which supplement each other."

【参考译文】

Ups and Downs of Conflicts between Democratic Transition and Political Violence: Taking Indonesia as an Example

×××

Abstract

Compared with "authoritarian rule in a steady period", the initial period of a transition to democracy is a historical stage characterized by increasing risk of political violence. This paper points out that although democratization contains a logical mechanism of violence intensification and violence alleviation, political trust in the early period of democratization is particularly hard to come by in a society with ethnic divisions and therefore, the mechanism of violence intensification often appears earlier than the mechanism of violence alleviation. The question is why does the level of violent conflict in different countries experiencing social disruption still show obvious differences during the transitional period? Namely, why is the level of violent conflict in some countries higher than that in other countries? This study focuses on "political tolerance" as a way of explaining the differences in multiethnic social transformation levels of violence. In this paper, political tolerance is understood as having two complementary dimensions — "mass culture" and "awareness of elites", which supplement each other. Simply stated, places with a higher level of political tolerance have a lower level of violent conflict, and vice versa. This paper presents and explains its core argument by taking Indonesia as an example.

【例 2】

集体决策的政治与协议的政治
——解决冲突的两种公平方式

×××

摘要

　　政治是对人们之间冲突的解决。一个理想的社会是不存在冲突的社会,但当冲突不可避免时,政治文明的进步就表现为不断寻求更公平的冲突解决方式。在现代条件下,这就表现为建立和完善集体决策的政治与协议的政治。同时,虽然在某些时候我们的行动可以不考虑公平的要求,但这些行动必须以将我们与冲突者的关系重新纳入公平框架之中为目的。公平意味着冲突中的各方不能自行充当其他人的裁决者,而必须让制度来裁决他们间的冲突,集体决策的政治与协议的政治就是建立这种制度的两种方式。

　　　　　　　　　　(张乾友. 集体决策的政治与协议的政治:解决冲突的两种公平方式[J].
　　　　　　　　　　行政论坛,2017(2):12-19.)

1. 词汇层

集体决策 group decision-making;协议的政治 politics of agreement;解决 resolve;冲突 conflict;建立和完善 establish and perfect;公平框架 fair framework;裁决者 adjudicator。

2. 句子层

这篇摘要相对于上一篇来说简单一些,所以翻译起来困难较小。

E. g.　一个理想的社会是不存在冲突的社会,但当冲突不可避免时,政治文明的进步就表现为不断寻求更公平的冲突解决方式。

本句中信息之间的关系有明确显示——"但"和"当……时",所以直译为英语即可。本句翻译困难在于"不断寻求更公平的冲突解决方式"中逻辑关系的明晰上。笔者将动态的汉语"不断寻求"理解为静态的英语"in the continuous efforts",而"更公平的冲突解决方式"虽然字面表示"寻求"的对象,按英语信息的逻辑关系,实则表示"寻求"努力的目的。因此,本句可译为：“An ideal society has no conflicts; however, when conflicts are unavoidable, the progress of a political civilization shows itself in the continuous efforts which are made to find equitable solutions."

E. g. 同时，虽然在某些时候我们的行动可以不考虑公平的要求，但这些行动必须以将我们与冲突者的关系重新纳入公平框架之中为目的。

笔者认为完全可以删除本句开头"同时"二字，紧随其后的"虽然……"，两者连接起来读来怪异（疑似原文写作有瑕疵，翻译时可以不予理会）。对"不考虑公平的要求"理解要超越文本本身，因为虽然用"考虑"作用于"公平"，但是地道的英文表述是"meet the requirements for..."，切不可望文生义。"这些行动必须以将我们与冲突者的关系重新纳入公平框架之中为目的"中"以……为目的"可转换为"aim to do sth."，"将……纳入……"可转换为"integrate ... into..."，这种转换彰显词汇层的基本功。本句可译为："Although on occasion we do not have to meet the requirements for fairness when we act, we must aim to integrate the relationship between us and our opponent(s) into a fair framework."

【参考译文】

Politics of Group Decision-making and Politics of Agreement
—— Two Fair Ways to Resolve Conflicts
×××

Abstract

Politics is a way of resolving conflicts among people. An ideal society has no conflicts; however, when conflicts are unavoidable, the progress of a political civilization shows itself in the continuous efforts which are made to find equitable solutions. Under modern conditions, this is seen in the establishing and perfecting of the politics of group decision-making and agreement. Although on occasion we do not have to meet the requirements for fairness when we act, we must aim to integrate the relationship between us and our opponent(s) into a fair framework. Fairness means that no party involved in a conflict, shall itself act as an adjudicator for any other party. It is the political system that adjudicates the conflict between them. The politics of group decision-making and agreement are two ways to establish such a system.

练 习 题

一、将下列文本译为汉语。

摘要 1：

Aspiration in Polish: A Sound Change in Progress?
×××

Abstract

Aspiration in the production of /p/t/k/ before a stressed vowel has been traditionally believed to be typical for English, but not Polish. In the speech of Polish-English bilinguals, the length of the Voice Onset Time in English and Polish has been found to be conditioned by phonetic universal effects as well as the stylistic and attitudinal factors, with aspiration functioning as one of the markers. Recently, the use of aspiration in Polish seems to be growing not as a sign of English-accented speech but rather an element of emphatic style. The study reported here explores the use of the longer VOT values in Polish from the perspective of stylistic conditioning, adopting the attention-to-speech paradigm and discussing the relationship between the observed variability and the predictions formed on the basis of universal phonetic tendencies. The observed tendency to lengthen the VOT values in the universally lengthening contexts is claimed to be further affected by the attention paid to speech, which suggests a possible prestigious function of aspiration in Polish. The intriguing questions emerging from the study are whether the change is the effect of language experience and whether it spreads from bilingual/advanced learners of English to mono-lingual speakers.

(WANIEK – KLIMCZAK E. Aspiration in Polish: A Sound Change in Progress? [C]//PAWLAK M, BIELAK J. New Perspectives in Language, Discourse and Translation Studies. Berlin Heidelberg: Springer, 2011:3 – 11.)

摘要 2:

The Application of Text Type in Non-literary Translation Teaching

×××

Abstract

Despite the relevance of text type to translation practice, especially to Chinese-English non-literary translation in which the two languages display remarkable textual differences, there has been a general lack of attention to text type related issues in translation teaching in Chinese universities. Centered upon the translation of literary works, translator training in China has long focused on techniques at the lexical and syntactic level, and a text-based approach has yet to be adopted. This coincides with the clear tendency in the assignment of non-literary translation when students are quite active in making adaptation at the lexical and syntactic level but much more reluctant to make decisions at the textual level. Despite their intuitive awareness of the textual differences between the two languages, they are not well trained to effectively deal with such differences so that the translated text can fulfill its communicative function. This article is an attempt to pinpoint this problem and highlight the necessity of including text type in translation pedagogy. It also proposes a new teaching framework within which text type is taught in a systematic manner.

(CHEN Q J. The Application of Text Type in Non – literary Translation Teaching[J]. Translation & Interpreting Studies,2010, 5(2):208 – 219.)

二、将下列文本译为英语。

摘要1：

高校教师计算机自我效能感与计算机态度的实证研究

×××

摘要

  已有研究表明，计算机自我效能感与计算机态度是影响高校教师进行信息技术与课程整合的重要变量，因此非常有必要对教师的计算机自我效能感与计算机态度之间的关系进行探讨。研究选取山东省4所高校教师为调查对象，对高校教师的计算机自我效能感与计算机态度各个维度之间的相关性进行了分析。结果显示：高校教师的计算机自我效能感与他们的计算机态度各个维度以及总和之间存在着统计意义上的相关；计算机态度中的可感知控制以及行为因素两个维度对于教师计算机自我效能的预测作用最大。

  （孙先洪.高校教师计算机自我效能感与计算机态度的实证研究[J].江苏教育，2017(5)：64-66.）

摘要2：

评价理论对语篇翻译的启示

×××

摘要

  本文研究评价理论对翻译研究的贡献。作为系统功能语言学在基调理论方面的发展，评价理论重点研究语篇中的表态资源、态度来源及语篇姿态。我们的研究表明，翻译理论历来重视对于语篇中态度意义的研究，但缺乏可操作的工具。我们认为，评价理论可以作为翻译中的态度分析工具。此外，评价理论对于语篇中的对话性研究，对于翻译研究也颇有启示。将评价理论引入翻译研究将会推动翻译理论与翻译教学的发展。

  （张先刚.评价理论对语篇翻译的启示[J].外语教学，2007，28(6)：33-36.）

第 12 章　计算机与网络技术语篇翻译

12.1　计算机与网络技术语篇的特点

随着计算机与网络技术的不断发展,各种报道层出不穷,产生了很多与之有关的科技或科普文章。这些文章记载了计算机与网络技术的发展历程和最新发展动态,为不同人群提供及时的信息帮助,用以解决社会生活中形形色色的实际问题,在推动社会进步的同时,也不断促进计算机与网络技术不断向前发展。

计算机与网络技术语篇常见的特点如下:

(1)语言风格平实质朴,几乎无修饰和渲染,无华丽的辞藻。

(2)专业词汇多,词汇隐喻化(词汇来自另一领域,如木马程序 Trojan horse、计算机病毒 computer virus、宽带网络 broadband networks 等)明显,缩略词多(往往是为了写作时节约篇幅或口头交流时提高效率)。

(3)信息性强,逻辑推进自然、可信(技术支撑明显,图表呈现清晰),表述客观,以理服人等。

12.2　计算机与网络技术语篇翻译实例

12.2.1　英译汉

Net neutrality: why are Americans so worried about it being scrapped?

Most of the world won't be affected by the changes, so are they a problem? No, if you are a tech monopoly — but yes if you don't want a two-tier internet.

Net neutrality is the idea that internet service providers should not interfere in the information they transmit to consumers.

Photograph：UPI /Barcroft Images

AjitPai，head of the US telecoms regulator，revealed sweeping changes on Tuesday to overturn rules designed to protect an open internet.

The regulations，put in place by the Obama administration in 2015，enshrined the principle of "net neutrality" in US law. Net neutrality is the idea that internet service providers should not interfere in the information they transmit to consumers，but should instead simply act as "dumb pipes" that treat all uses，from streaming video to sending tweets，interchangeably.

Net neutrality is unpopular withInternet service providers（ISPs），who struggle to differentiate themselves in a world where all they can offer are faster speeds or higher bandwidth caps，and who have been leading the push to abandon the regulations in the US.

On the other side of the battle are companies relying on the Internet to connect to customers. Their fear is that in an unregulated Internet，ISPs may charge customers extra to visit certain websites，demand fees from the sites themselves to be delivered at full-speed，or privilege their own services over those of competitors.

The fear is well-founded. Outside the US，where net neutrality laws are weaker and rarely enforced，ISPs have been experimenting with the sorts offavouritism that a low-regulation environment permits.

In Portugal，mobile carrier MEO offers regular data packages，but it also offers，for € 4.99 a month，10GB "Smart Net" packages. One such package for video provides 10GB of data exclusively for YouTube，Netflix，Periscope and Twitch，while one for messaging bundles six apps including Skype，WhatsApp and FaceTime.

In New Zealand，Vodafone offers a similar service：for a daily，weekly，or monthly fee，users can exempt bundles of apps from their monthly cap. A "Social Pass" offers unlimited Facebook，Instagram，Snapchat and Twitter for NZ$10 for 28 days，while a "Video Pass" gives five streaming services — including Netflix but not YouTube — for $20 a month...

(https：//www.theguardian.com/technology/2017/nov/22/net-neutrality-internet-why-americans-so-worried-about-it-being-scrapped)

本节选文讲述了美国政府、美国民众和互联网服务提供商等机构对废除网络中立表达关切的原因。

1. 词汇层

Net neutrality 网络中立；scrap 废弃；tech monopoly 技术垄断；a two-tier internet 二级互联网；sweeping change 全面改变；the Obama administration 奥巴马政府；enshrine 把……奉为神圣；bandwidth cap 带宽上限；well-founded 理由充足的；favouritism 偏爱；data package【计】数据包；Portugal 葡萄牙；mobile carrier 移动运营商；bundle 捆绑；streaming service 流媒体服务。

2. 句子层

E. g. Most of the world won't be affected by the changes, so are they a problem? No, if you are a tech monopoly — but yes if you don't want a two-tier internet.

相对而言,汉语多用主动语态表意用,中、高级英语(尤其是科技英语)多用被动语态表意。英译汉时,为使译文读来顺畅,须尽可能将英语的被动语态转换为主动语态。此处的"No"和"yes"是对"so are they a problem?"的答语,如果仅仅翻译为"不是"和"是的",汉语将读来不是很流畅,也不太符合书面汉语的严肃风格。考虑到译文的阅读效果,笔者将"No"和"yes"分别译为一个句子。本句可译为:"废除网络中立不会对全球大部分区域造成影响,那么废除网络中立是否会产生问题?如果您同意技术垄断,废除网络中立不会产生问题;反之,如果您不想使用二级互联网,则会认为废除网络中立肯定会产生问题。"

E. g. AjitPai, head of the US telecoms regulator, revealed sweeping changes on Tuesday to overturn rules designed to protect an open internet.

外国人名的翻译须遵循专有名词译法,须与官方保持一致。名人的人名因读者耳熟能详,一般直接用官方的汉语译法,不必注出原名。如果非名人或词典和官方未给出译法,一般采用音译法,译为中性的汉字。比如这句中的"AjitPai",笔者将其译作"阿基特帕伊"。"head of the US telecoms regulator"为本句主语"AjitPai"的同位语。一般情况下,若同位语较短,可译作前置定语;若较长,可译作一个汉语单句。本句可译为:"周二,美国电信监管部门负责人阿基特·帕伊(AjitPai)公布了全面改革措施,旨在颠覆为保护互联网开放性而制定的规则。"

E. g. The regulations, put in place by the Obama administration in 2015, enshrined the principle of "net neutrality" in US law.

本句主语虽为"The regulations"(物称主语),但笔者将本句译为"2015年奥巴马政府颁布相应法规,在美国法律中明文确立'网络中立'原则。",将主语改为"奥巴马政府"(虽然字面为"政府",实则指人),因为相对于英语而言,汉语多用人称表意。

E. g. Net neutrality is unpopular with Internet service providers (ISPs), who struggle to differentiate themselves in a world where all they can offer are faster speeds or higher bandwidth caps, and who have been leading the push to abandon the regulations in the US.

如同上一例一样,本句的主语也是物"Net neutrality","Internet service providers"后跟了三个定语从句,第一和第三个均用"who"引导的定语从句修饰"Internet service providers",而第二个用"where"引导的从句修饰其前的"world",属于第一个定语从句中套用的从句。英译汉时,当构成定语从句的字眼大于8个时,一般要断为汉语单句。由此可见,翻译时译者要居高临下,熟悉原文整体与局部性信息及其相互关系,要心怀英汉差异,做到妥当转换,译文才会字正句顺。本句可译为:"互联网服务提供商(ISPs)并不认可'网络中立'理念。在充满竞争的互联网服务领域中,互联网服务提供商需努力通过提供更快的网速和更高的带宽上限,从同行中脱颖而出。因此,互联网服务提供商一直率先努力说服美国政府取消'网络中立'法案。"

E. g. Their fear is that in an unregulated Internet, ISPs may charge customers extra to visit certain websites, demand fees from the sites themselves to be delivered at full-speed, or privilege their own services over those of competitors.

译文的通顺与否固然受多种因素影响。除了逻辑、句式搭建、文化认知度等大的方面外，更多的是词汇层细节方面的处理。本句句首的"Their fear"按照字面意思译为"他们的担心是……"不如译为"这些公司担心……"好，原因有二：① 汉语多重复，英语多替代（"their"指代上一句提到的"companies"，是避免重复的做法）；② 英语是静态语言，而汉语是动态语言。原文中名词"fear"转换为动词"担心"正是汉英在语言上的动静转换的结果。本句可译为："这些公司担心在不受管制的互联网世界中，互联网服务提供商可能会向特定网站用户收取额外的费用，并通过设定用户浏览网站速度，收取全速浏览网站的费用，或要求网站提供比竞争对手更好的特权服务。"

【参考译文】

网络中立：为什么美国人如此担心废除"网络中立"？

废除网络中立不会对全球大部分区域造成影响，那么废除网络中立是否会产生问题？如果您同意技术垄断，废除网络中立不会产生问题；反之，如果您不想使用二级互联网，则废除网络中立肯定会产生问题。

（图下文字：网络中立的理念是防止运营商干预向客户传输的任何信息，保证网络数据传输的"中立性"。摄影：UPI／巴克罗夫特）

周二，美国电信监管部门负责人阿基特帕伊（AjitPai）公布了全面改革措施，旨在颠覆为保护互联网开放性而制定的规则。

2015年奥巴马政府颁布相应法规，在美国法律中明文确立"网络中立"原则。"网络中立"的理念是互联网服务提供商不应干预向客户传输的所有信息，而应简单作为"无声管道"提供所有互联网服务，如视频流传输、推特信息发送以及信息互换等。

互联网服务提供商（ISPs）并不认可"网络中立"理念。在充满竞争的互联网服务领域中，互联网服务提供商需努力通过提供更快的网速和更高的带宽上限，从同行中脱颖而出。因此，互联网服务提供商一直率先努力说服美国政府取消"网络中立"法案。

这场博弈的另一方是依靠互联网与客户联系的公司。这些公司担心在不受管制的互联网世界中，互联网服务提供商可能会向特定网站用户收取额外的费用，并通过设定用户浏览网站速度，收取全速浏览网站的费用，或要求网站提供比竞争对手更好的特权服务。

上述担心理由充足。美国之外的区域均处于网络中立法律不健全且很少被强制执行的状态，互联网服务提供商一直在低调地享受着低监管环境向其提供的种种偏袒服务。

在葡萄牙，移动运营商MEO不但向用户提供常规套餐服务，还向用户提供每月4.99欧元的10GB"智能网络"套餐服务。在视频套餐服务中，MEO向用户提供YouTube、Netflix、Periscope和Twitch专用的10GB视频套餐；而在另一流量套餐中，MEO强行捆绑了包括Skype、WhatsApp和FaceTime在内的六款软件。

在新西兰，沃达丰公司也向用户提供类似的套餐服务：通过定期（每日、每周或每月）缴纳费用，用户使用捆绑的应用软件时不受月度流量上限的限制。其中，名为"社交通行证"的服务可使用户无限量使用Facebook、Instagram、Snapchat和Twitter软件28天，每月只需缴纳10新西兰元。然而，名为"视频通行证"的服务可使用户每月只需缴纳20新西兰元，就可以使用五种流媒体服务（包括 Netflix，但不包括 YouTube）。

12.2.2 汉译英

微星 Z170 GAMING M7 主板怎么样？

2015-10-08 11:10 | 作者：电脑维修技术网 | 来源： | 参与评论 | 点击：7384 次

　　距离 Intel 发布 Skylake 构架的处理器已有近两个月的时间，除了入门级的 H110 平台，各个板卡大厂的 100 系芯片组主板产品线都已备全。占据了大半江山的 B150 性价比出众，最受广大玩家欢迎，售价为五百到一千元不等。这其中千元级别的产品虽然用料豪华、功能如八仙过海，但是毕竟不是 Z170 的高端平台。相信追求极致性能的玩家一定不想给自己或者别人任何挑剔的理由。

　　微星 Z170 GAMING M7 主板就是一款无可挑剔的主板。

微星游戏王者系列主板官方介绍

　　游戏的主宰：想要成为游戏大师，你需要一套最顶级的系统。为呼应玩家成为游戏大师的渴望，微星推出游戏王者系列主板，并在产品型号的后缀上加入了代表 Master（大师）的'M'这个符号说明微星王者系列主板在拥有十分杰出设计的同时，也能协助你的游戏表现更上一层楼。

微星 Z170 GAMING M7 主板的提升

　　我们先来看一下原生的 Z170 芯片组相对前一代产品具有怎样的提升。

一、中央处理器插槽由原来的 LGA 1150 变为 LGA 1151

　　鉴于本次 100 系芯片组升级为整个平台带来了不少惊喜的提升，就先不要抱怨每次升级 CPU 插槽都不兼容上一代产品的设计了。就像产品经理常说的那样，我们要保证有足够的利益进行产品的售后服务和研发环节。从全球发展的角度来看，确实如此。

二、高贵的 DDR4 内存

由于 Skylake 内部集成了双内存控制器，支持 DDR3L 和 DDR4 内存。而上一代也只有顶级的 X99 平台可以支持 DDR4 内存。DDR4 内存金手指部分的"豁口"，也就是防呆口是要比 DDR3 的更靠近中央，还有了一定程度的弯曲。所以具体新 100 系列芯片组平台需要用到哪种内存完全是看厂商了。目前市面上的主流 100 系列平台多是只支持 DDR4 的。而作为 100 系芯片组目前最高端的 Z170，厂商只配 DDR3 内存插槽是没有理由的。

三、DMI 2.0 总线升级到 DMI 3.0

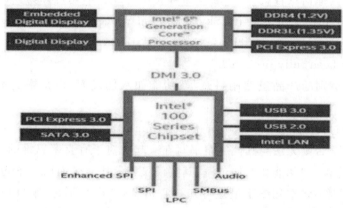

由于新的 100 系芯片组原生支持 PCI-E 3.0，DMI 2.0 总线升级为 DMI 3.0。得益于带宽的提升，USB 3.1、SATA-E 还有 M.2 这些设备的传输速度瓶颈被轻松突破。Z170 最高支持 3 个 SATA-Express 接口。虽然笔者从未在存储方面搞出些牛逼的名堂，不过看样子喜欢搭阵列的朋友们有福了。

以上三点是原生 100 系芯片组相对于上一代产品的改变和提升，Z170 芯片组官方规格设置最高。

（http://www.pc811.com/tuijian/26614.html）

这是 4 年前互联网上一篇文章的节选部分，对微星 Z170 GAMING M7 主板提升做了较详细的介绍。

该网文翻译困难较大，表现在文章本身较为专业、术语较多和译文英语搭建较困难等方面。翻译时，译者要"眼观六路"，彻底弄清原文意思，给出准确的专业词汇，合理耐心搭建英语句式，清晰传递出原文意思。

1. 词汇层

Skylake 构架的处理器 Skylake-based processor；入门级的 entry-level；芯片组主板产品

线 chipset main board product line；性价比 cost performance；性价比高 cost-effective；主板 mainboard；无可挑剔的 flawless；游戏王者系列 game king series；中央处理器插槽 CPU socket；不兼容 incompatible with；研发 research and development；集成 integration；双内存控制器 dual memory controller；金手指部分的"豁口"the "opening" at the gold finger part；内存插槽 memory slot；总线 bus；传输速度瓶颈 transmission bottleneck；搭阵列 build array；规格 specification。

2. 句子层

E.g. 距离 Intel 发布 Skylake 构架的处理器已有近两个月的时间，除了入门级的 H110 平台，各个板卡大厂的 100 系芯片组主板产品线都已备全。

翻译本句前，须先理清信息层次。根据意群，翻译时宜从"除了……"之前逗号处断开，前后译为两句。第一句"距离……已有近两个月的时间"是汉语典型的写法——"先叙事，后表态"，译为英语时该转换为英语典型的写法——"先表态，后叙事"，译为"It has been nearly two months since..."。第二句翻译时要抓住核心概念"100 系芯片组主板产品线"，将其译为第二句的主语，其他信息各就各位，翻译即可大功告成。本句可译为："It has been nearly two months since Intel released its Skylake-based processor. Except for the entry-level H110 platform，the 100-Series chipset main board product line of each major board manufacturer has been fully prepared."

E.g. 这其中千元级别的产品虽然用料豪华、功能如八仙过海，但是毕竟不是 Z170 的高端平台。

本句中"八仙过海"不可直译为"The Eight Immortals Crossing the Sea (a fairy tale in Chinese history)"，否则无法与其他字眼书写在一起。对于有直率思维的英语读者而言，这样做有很大的误导性。"功能如八仙过海"其实仅仅表示功能多而已。本句翻译目的语英语句式搭建并不难，因为将本句缩至最短可知核心信息是"产品不是（来自）平台。"，将其译作英语主句，其他问题即可迎刃而解。本句可译为："Though they function well and are made from high-quality materials，the products priced around ￥1,000 are not，in the final analysis，from high-end platforms like Z170."

E.g. 为呼应玩家成为游戏大师的渴望，微星推出游戏王者系列主板，并在产品型号的后缀上加入了代表 Master(大师)的 M 这个符号说明微星王者系列主板在拥有十分杰出设计的同时，也能协助你的游戏表现更上一层楼。

"推出"可用英语"launch"表示。本句的翻译困难在于如何明晰"微星推出游戏王者系列主板"后信息的关系。汉语为意合语言，逻辑关系词一般很少明示，信息呈竹节状不断推进；英语表意时讲求信息的层次感，主次分明，信息间是从属还是并列关系、并列关系中是承接、递进还是转折等，都一目了然。本句可译为："In response to players' desire to become game masters，MSI has launched the game king series mainboard with the suffix "M" (meaning "master") on the product models. This symbol indicates that MSI game king series mainboards not only have an outstanding design，but can also help players raise their game performance level." 从译文中可以看出，"并在产品型号的后缀上加入了代表 Master（大师）的 M 这个符号"以"with"短语的形式附着在前一个主谓结构之上。为表意清晰起

见,笔者将"说明……"译为另一英语句。

E. g. 就像产品经理常说的那样,我们要保证有足够的利益进行产品的售后服务和研发环节。从全球发展的角度来看,确实如此。

  本句的翻译困难在于"……利益进行产品的售后服务和研发环节"信息之间关系的挖掘上。相对于英语而言,汉语表意有时会模糊一些。字面来看,"利益"与"售后服务"和"研发环节"是"进行"的关系,该如何翻译"进行"才能保证这些信息写为一句英语呢?笔者将其理解为:"利益"实为"利润"(profit),准备足够的利润以"提高售后服务并推动研发工作"。因此,当我们准确理解"进行"在这个上下文中的具体意思时,译文生成就变得简单了。"从全球发展的角度来看,确实如此。"是对上一句的肯定评价,译为用"which"引导的定语从句,和上一句连为一句,将句子拉长。本句可译为:"As the product manager often says, "We want to ensure that there is enough profit reserved to improve after-sale services of our products and promote research and development work, which is true from the perspective of global development.""

E. g. DDR4 内存金手指部分的"豁口",也就是防呆口是要比 DDR3 的更靠近中央,还有了一定程度的弯曲。

  此处"金手指""豁口"和"防呆口"要分别直译为"the gold finger""opening"和"the fool-proof interface",这是计算机与网络技术术语中众多隐喻化词汇中的 3 个例子。

  "也就是"表明其前后信息呈并列状态,其后的"更靠近中央"和"还有了一定程度的弯曲"和主语"'豁口'"可形成两个主谓结构。很多汉语句子中都存在两个以上的主谓结构,不可将这些主谓结构统统译作一个个的英语单句,否则译文英语的可读性很低。好的英语句子往往以 1~2 个主谓结构为中心(译作主句),让其他主谓结构设法附着在作主句的主谓结构之上。本句中的两个主谓结构即可这样处理。笔者用"更靠近中央"作主句的谓语,将"还有了一定程度的弯曲"处理为从句,英语译文信息的层次感跃然纸上。本句可译为:"The 'opening' at the gold finger part of the DDR4 memory, namely, the fool-proof interface, which has a certain degree of bending, is closer to the center than that of the DDR3."

E. g. 以上三点是原生 100 系芯片组相对于上一代产品的改变和提升,Z170 芯片组官方规格设置最高。

  翻译本句时要注意细节之处的处理。逗号前后 2 个主谓结构无从属关系,要处理为并列关系;将"以上三点是原生 100 系芯片组相对于上一代产品的改变和提升"缩至最短,即是"三点是改变和提升"。其中,将"是"译作"involve"要比译作"are"语气正式很多;"改变和提升"前的定语较长,但逻辑关系很清楚:"原生 100 系芯片组"与"改变和提升"是所有关系,用"of"表示,而"相对于上一代产品"可译作"compared with the previous generation of chipsets"。因此,本句可译为:"The above three points involve changes and improvements in the original 100-Series chipset as compared with the previous generation of chipsets; the official standard specification of the ZI70 chipset is the highest of all."

【参考译文】

How About the Mainboard, MSI Z170 GAMING M7?

11:10, October 8, 2015 | Author: www.pc811.com | Source:
| Comments | Clicks: 7,384 times

It has been nearly two months since Intel released its Skylake-based processor. Except for the entry-level H110 platform, the 100-Series chipset main board product line of each major board manufacturer has been fully prepared. B150, which accounts for more than half of the market, has an excellent cost performance record and is the most popular brand with the majority of players. Moreover, it is cost-effective, selling for prices ranging from ￥500 to 1,000. Though they function well and are made from high-quality materials, the products priced around ￥1,000 are not, in the final analysis, from high-end platforms like Z170. It is believed that the players in pursuit of ultimate performance do not want to give themselves or others any reason to be picky.

The Mainboard, MSI Z170 GAMING M7, is flawless.

Official introduction to MSI game king series mainboards

Game master: To become a game master, you need a top-of-the-line system. In response to players' desire to become game masters, MSI has launched the game king series mainboard with the suffix "M" (meaning "master") on the product models. This symbol indicates that MSI gameking series mainboards not only have an outstanding design, but can also help players raise their game performance level.

The upgraded MSI Z170 GAMING M7 mainboard

Let's first look at how the original upgraded Z170 chipsets compared with previous generations of products.

Ⅰ. The CPU socket has been changed from the original LGA 1150 to LGA 1151

Since this 100-Series chipset upgrade has brought a lot of surprises with respect to the entire platform, please do not complain that every upgrading of the CPU socket is incompatible with the product design of the previous generation. As the product manager often says, "We want to ensure that there is enough profit reserved to improve after-sale services of our products and promote research and development work, which is true from the perspective of global development.

Ⅱ. "Noble" DDR4 memory

Due to the integration of a dual memory controller in Skylake, both the DDR3L and DDR4 memory bars are supportable, while the previous generation of memory bars is compatible only with X99, the top-level platform. The "opening" at the gold finger part of the DDR4 memory, namely, the fool-proof interface, is closer to the center than that of the DDR3, which has a certain degree of bending. Therefore, the kind of memory bar to be used for the new 100-Series chipset platform depends entirely on the manufacturer. At present, most of the mainstream 100-Series platforms on the market support only DDR4. As for Z170, the high end product of the 100-Series chipset, there is really no justification for manufacturers having configured the DDR3 memory slot only.

Ⅲ. DMI 2.0 The computer bus has been upgraded to DMI 3.0

Since the new original 100-Series chipset supports PCI-E 3.0, the DMI 2.0, the bus has been upgraded to DMI 3.0. Thanks to the boost in bandwidth, USB 3.1, SATA-E, and M.2 have easily broken through the transmission bottleneck of these devices. Z170 supports up to 3 SATA-Express interfaces. Although the author has not invented (developed) anything notable in the area of data storage, it is good news for those friends who like to build arrays.

The above three points involve changes and improvements in the original 100-Series chipset as compared with the previous generation of chipsets; the official standard specification of the ZI70 chipset is the highest of all.

练 习 题

一、请将下列文本译为汉语。

Properties

Computer networking may be considered a branch of electrical engineering, electronics engineering, telecommunications, computer science, information technology or computer engineering, since it relies upon the theoretical and practical application of the related disciplines.

A computer network facilitates interpersonal communications allowing users to communicate efficiently and easily via various means: email, instant messaging, online chat, telephone, video telephone calls, and video conferencing. A network allows sharing of network and computingresources. Users may access and use resources provided by devices on the network, such as printing a document on a shared network printer or use of a shared storage device. A network allows sharing of files, data, and other types of information giving authorized users the ability to access information stored on other computers on the network. Distributed computing uses computing resources across a network to accomplish tasks.

A computer network may be used by security hackers to deploy computer viruses or computer worms on devices connected to the network, or to prevent these devices from accessing the network via a denial-of-service attack.

Network packet

Computer communication links that do not support packets, such as traditional point-to-point telecommunication links, simply transmit data as a bit stream. However, most information in computer networks is carried in packets. A network packet is a formatted unit of data (a list of bits or bytes, usually a few tens of bytes to a few kilobytes long) carried by a packet-switched network. Packets are sent through the network to their destination. Once the packets arrive they are reassembled into their original message.

Packets consist of two kinds of data: control information, and user data (payload). The control information provides data the network needs to deliver the user data, for example: source and destination network addresses, error detection codes, and sequencing information. Typically, control information is found in packet headers and trailers, with payload data in between.

With packets, the bandwidth of the transmission medium can be better shared among users than if the network were circuit switched. When one user is not sending packets, the link can be filled with packets from other users, and so the cost can be shared, with relatively little interference, provided the link isn't overused. Often the route a packet needs to take through a network is not immediately available. In that case the packet is queued and waits until a link is free...

(https://en.wikipedia.org/wiki/Computer_network)

二、请将下列文本译为英语。

计算机系统

简介

计算机系统是按人的要求接收和存储信息,自动进行数据处理和计算,并输出结果信息的机器系统。计算机是脑力的延伸和扩充,是近代科学的重大成就之一。

计算机系统由硬件(子)系统和软件(子)系统组成。前者是借助电、磁、光、机械等原理构成的各种物理部件的有机组合,是系统赖以工作的实体。后者是各种程序和文件,用于指挥全系统按指定的要求进行工作。

自1946年第一台电子计算机问世以来,计算机技术在元件器件、硬件系统结构、软件系统、应用等方面,均有惊人进步,现代计算机系统小到微型计算机和个人计算机,大到巨型计算机及其网络,形态、特性多种多样,已广泛用于科学计算、事务处理和过程控制,日益深入社会各个领域,对社会的进步产生深刻影响。

电子计算机分数字和模拟两类。通常所说的计算机均指数字计算机,其运算处理的数据,是用离散数字量表示的。而模拟计算机运算处理的数据是用连续模拟量表示的。模拟机和数字机相比较,其速度快、与物理设备接口简单,但精度低、使用困难、稳定性和可靠性差、价格昂贵。故模拟机已趋淘汰,仅在要求响应速度快,但精度低的场合尚有应用。把二者优点巧妙结合而构成的混合型计算机,尚有一定的生命力。……

(https://baike.so.com/doc/5912132-6125040.html)

第 13 章　旅游语篇翻译

13.1　旅游语篇的分类及特点

13.1.1　旅游语篇的分类

旅游活动,是"人们为了休闲、娱乐、探亲访友或者公务目的而进行的非定居性旅行和在游览过程中所发生的一切关系和现象的总和"(陈刚,2014)。可以这样说,凡是记录旅游活动的文章均可视作旅游语篇。

旅游语篇大致包括以下方面:
(1)景点介绍;
(2)风景明信片;
(3)景区画册;
(4)旅游杂志;
(5)旅游地图;
(6)景区广告标语;
(7)游客须知;
(8)古迹楹联解说词。

故此,凡是以旅游活动有关文本为对象的翻译都可视作旅游语篇翻译。

比如以下这些图片(见图 13-1~图 13-4),我们并不陌生(图片来自"百度图片")。

图 13-1　图片示例(一)

图 13-2　图片示例(二)

图 13-3　图片示例(三)

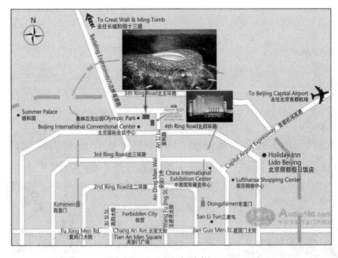

图 13-4　图片示例(四)

13.1.2　旅游语篇的特点

(1)在内容上,文化内涵丰富。

党的十九大报告指出,"文化是一个国家、一个民族的灵魂"。通过考古挖掘和典藏文献梳理提炼,我国已开发了一大批旅游景点。厚重的文化内涵提升了旅游品质,加深了游客的旅游体验,丰富了游客的精神和文化生活。在这方面,中外相同。

(2) 在语言风格上，英语往往表述简约，描述客观，平实易懂；汉语往往辞藻华丽，修辞手法多，抒情风格明显，引语（诗词歌赋、名人题词等）多。

(3) 文本类型包罗万象。

在翻译实践中，旅游语篇在翻译时除尽可能保持原始信息外，还要对语言风格进行调整，保证译文信息的连贯性。

13.2 旅游语篇翻译实例

13.2.1 英译汉

The Grand Canyon

The Grand Canyon is a river valley in the Colorado Plateau that exposes uplifted Proterozoic and Paleozoic strata, and is also one of the six distinct physiographic sections of the Colorado Plateau province. It is not the deepest canyon in the world (KaliGandaki Gorge in Nepal is much deeper). However, the Grand Canyon is known for its visually overwhelming size and its intricate and colorful landscape. Geologically, it is significant because of the thick sequence of ancient rocks that are well preserved and exposed in the walls of the canyon. These rock layers record much of the early geologic history of the North American continent.

Uplift associated with mountain formation later moved these sediments thousands of feet upward and created the Colorado Plateau. The higher elevation has also resulted in greater precipitation in the Colorado River drainage area, but not enough to change the Grand Canyon area from being semi-arid. The uplift of the Colorado Plateau is uneven, and the Kaibab Plateau that Grand Canyon bisects is over a one thousand feet (300 m) higher at the North Rim (about 1,000 ft or 300 m) than at the South Rim. Almost all runoff from the North Rim (which also gets more rain and snow) flows toward the Grand Canyon, while much of the runoff on the plateau behind the South Rim flows away from the canyon (following the general tilt). The result is deeper and longer tributary washes and canyons on the north side and shorter and steeper side canyons on the south side.

Temperatures on the North Rim are generally lower than those on the South Rim because of the greater elevation (averaging 8,000 feet or 2,400 metres above sea level). Heavy rains are common on both rims during the summer months. Access to the North Rim via the primary route leading to the canyon (State Route 67) is limited during the winter season due to road closures.

The Grand Canyon is part of the Colorado River basin which has developed over the past 70 million years, in part based on apatite (U-Th)/He thermochronometry showing that Grand Canyon reached a depth near to the modern depth by 20 Ma. A recent study examining caves near Grand Canyon places their origins beginning about 17 million years ago. Previous estimates had placed the age of the canyon at 5~6 million years. The study, which was published in the journal *Science* in 2008, used uranium-lead dating to analyze calcite deposits found on the walls of nine caves throughout the canyon. There is a substantial amount of controversy because this research suggests such a substantial departure from prior widely supported scientific consensus. In December 2012, a study published in the journal *Science* claimed new tests had suggested the Grand Canyon could be as old as 70 million years. However, this study has been criticized by those who support the "young canyon" age of around six million years as "[an] attempt to push the interpretation of their new data to their limits without consideration of the whole range of other geologic data sets."

Weather in the Grand Canyon varies according to elevation. The forested rims are high enough to receive winter snowfall, but along the Colorado River in the Inner Gorge, temperatures are similar to those found in Tucson and other low elevation desert locations in Arizona. Conditions in the Grand Canyon region are generally dry, but substantial precipitation occurs twice annually, during seasonal pattern shifts in winter (when Pacific storms usually deliver widespread, moderate rain and high-elevation snow to the region from the west) and in late summer (due to the North American Monsoon, which delivers waves of moisture from the southeast, causing dramatic, localized thunderstorms fueled by the heat of the day). Average annual precipitation on the South Rim is less than 16 inches (41 cm), with 60 inches (150 cm) of snow; the higher North Rim usually receives 27 inches (69 cm) of moisture, with a typical snowfall of 144 inches (370 cm); and Phantom Ranch, far below the canyon's rims along the Colorado River at 2,500 feet (762 m) gets just 8 inches (20 cm) of rain, and snow is a rarity.

(来自"维基百科")

这是"维基百科"上介绍美国科罗拉多大峡谷的一篇网文，涉及科罗拉多大峡谷形成历史、地层结构、地质现状、天气（降雨和降雪量）等信息。本文文风朴素，表述客观，长、难句少，读来平实易懂。这是英语国家景点介绍文本常见的文风。

1. 词汇层

the Colorado Plateau 科罗拉多高原；Proterozoic 元古代的；Paleozoic strata 古生代地层；physiographic section 地形剖面；elevation 海拔；drainage area 流域；semi-arid 半干旱的；the Kaibab Plateau 凯巴布高原；bisect 一分为二，横贯；the South Rim 南缘；road closure 道路封闭；the Colorado River basin 科罗拉多河盆地；uranium-lead dating【地质】铀铅测年；calcite deposit 方解石沉积岩；Tucson 图森（美国亚利桑那州南部城市）；monsoon 季风。

2. 句子层

E.g. The Grand Canyon is a river valley in the Colorado Plateau that exposes uplifted Proterozoic and Paleozoic strata, and is also one of the six distinct physiographic sections of the Colorado Plateau province.

本句包含3个逻辑上的主谓结构，分别是"The Grand Canyon is a river valley""that exposes..."和"... is also one of..."，意思不难理解，困难在于如何处理"that exposes uplifted Proterozoic and Paleozoic strata"信息上。要译为前置定语修饰"the Colorado Plateau"吗?这个"that"引导的定语从句构成字眼虽未超过8个字,但若译为前置定语修饰"科罗拉多高原"后,与第一个主谓结构一起译为一句汉语的话,读来会相当不流畅。因此,笔者决定将该定语从句采用拆译法,译为一个汉语单句,这也迎合了汉语一般用短句表意的做法。本句可译为:"科罗拉多大峡谷是科罗拉多高原上的河谷,也是科罗拉多高原地区六大显著性地形剖面之一。因地壳上升的缘故,从谷壁可以观察到裸露的元古代地层和古生代地层。"

E. g. There is a substantial amount of controversy because this research suggests such a substantial departure from prior widely supported scientific consensus.

本句中信息呈因果关系。英译汉时按照前因后果的顺序罗列即可。"substantial departure from"为书面语,在此意思是"巨大差异"。本句可译为:"该研究结果与先前有关科罗拉多大峡谷形成年代的广泛科学共识之间存在巨大差异,因而引起广泛争议。"

E. g. The forested rims are high enough to receive winter snowfall, but along the Colorado River in the Inner Gorge, temperatures are similar to those found in Tucson and other low elevation desert locations in Arizona.

本句较长,但是不难理解意思。英译汉时,可考虑从"but"处前后断为两句汉语。本句可译为:"峡谷边缘森林地带海拔高,所以冬季降雪量充沛。沿峡谷内科罗拉多河谷底地带,温度与亚利桑那州图森区域和其他低海拔沙漠区域的温度相似。"

E. g. Conditions in the Grand Canyon region are generally dry, but substantial precipitation occurs twice annually, during seasonal pattern shifts in winter (when Pacific storms usually deliver widespread, moderate rain and high-elevation snow to the region from the west) and in late summer (due to the North American Monsoon, which delivers waves of moisture from the southeast, causing dramatic, localized thunderstorms fueled by the heat of the day).

这个句子因为括号中注释性信息的增加而显得长一些。英译汉时,译者可照猫画虎,依然将括号中信息译入括号中,以保持原文的风貌。本句可译为:"科罗拉多大峡谷通常处于干旱状态,但每年冬天季节交替期(太平洋风暴通常为该区域带来大范围的中度降雨,并为西侧高海拔区域带来降雪)和夏末时期(受北美季风影响,从东南方向吹来的潮湿气流可在高温天气条件下引起局部雷暴天气)存在两次明显的降水过程。"

【参考译文】

科罗拉多大峡谷

科罗拉多大峡谷是科罗拉多高原上的河谷,也是科罗拉多高原地区六大显著性地形剖面之一。因地壳上升的缘故,从谷壁可以观察到裸露的元古代地层和古生代地层。尽管这并非全球最深的峡谷(尼泊尔的卡利甘达基峡谷要深得多),但科罗拉多大峡谷因其可产生视觉性震撼的规模和复杂多彩的景观而闻名于世。由于人们从科罗拉多大峡谷的谷壁可观察到保存良好的古生代厚岩石层,因此,科罗拉多大峡谷在地质学上也具有重要意义。这些岩石层很好地记录了北美大陆早期的地质演变历史。

后期发生与造山运动相关的地壳上移使沉积物向上移动了数千英尺而形成了科罗拉多

高原。海拔增高导致科罗拉多河流域降雨量增多,但降雨量并不足以改变科罗拉多大峡谷半干旱性质。科罗拉大高原的隆起高度不均匀。对于被科罗拉多大峡谷一分为二的凯巴布高原而言,其北缘比南缘高(约 1 000 英尺或 300 米)。北缘(降雨量和降雪量均高于南缘)上几乎所有的径流都流向科罗拉多大峡谷,而南缘后侧高原上的绝大部分径流都随着倾斜的地形从科罗拉多大峡谷流出。因此,科罗拉多大峡谷北侧因受径流冲刷而变得更深更长,而南侧变得更短更陡。

由于海拔较高(平均海拔高度为 8 000 英尺或 2 400 米),北缘的温度通常低于南缘。夏季,北缘区域和南缘区域都会迎来暴雨天气。67 号洲际公路是去往科罗拉多大峡谷北缘的主要路线,但因冬季封路,人们很难由此进入北缘区域。

对已有 7 000 万年历史的科罗拉多河盆地来说,科罗拉多大峡谷是它的一部分,因为磷灰石的(U-Th)/He 热同位素测年分析显示,科罗拉多大峡谷很久以前的深度已接近当前的 20Ma。近期一项针对科罗拉多大峡谷附近洞穴的考查表明,科罗拉多大峡谷的形成时间可追溯至约 1 700 万年前(而过去的科学研究估计科罗拉多大峡谷的形成年代仅为 500 至 600 万年前)。该研究使用铀-铅测年法,对科罗拉多大峡谷内九个洞穴墙壁上的方解石沉积岩实施检测,研究结果已于 2008 年发表在《科学》杂志上。该研究结果与先前有关科罗拉多大峡谷形成年代的广泛科学共识之间存在巨大差异,因而引起广泛争议。2012 年 12 月,《科学》杂志刊登的一篇研究报告称,最新测试结果表明科罗拉多大峡谷可能形成于 7 000 万年前。但该项研究受到"年轻峡谷"支持者(认为该峡谷的年龄约 600 万年)的强烈批评,并称此类研究"试图将科罗拉多大峡谷年龄解释向极限方向发展,而未考虑其他相关的地质学证据。"

科罗拉多大峡谷的天气随海拔高度的不同而变化。峡谷边缘森林地带海拔高,所以冬季降雪量充沛。沿峡谷内科罗拉多河谷底地带,温度与亚利桑那州图森区域和其他低海拔沙漠区域的温度相似。科罗拉多大峡谷通常处于干旱状态,但每年冬天季节交替期(太平洋风暴通常为该区域带来大范围的中度降雨,并为西侧高海拔区域带来降雪)和夏末时期(受北美季风影响,从东南方向吹来的潮湿气流可在高温天气条件下引起局部雷暴天气)存在两次明显的降水过程。科罗拉多大峡谷南缘的年平均降雨量不足 16 英寸(41 厘米),降雪量可达 60 英寸(150 厘米);地势较高的北缘年平均降雨量为 27 英寸(69 厘米),降雪量可达 144 英寸(370 厘米)。沿着科罗拉多河形成的"幻影牧场(Phantom Ranch)"区域海拔高度仅为 2 500 英尺(762 米),远低于科罗拉多大峡谷两缘,年平均降雨量为 8 英寸(20 厘米),降雪是极其罕见的。

13.2.2 汉译英

【例 1】

泰山古称岱山,又称岱宗。位于山东省中部,为中国五岳(泰山、华山、衡山、嵩山、恒山)之一。因地处东部,故称东岳。泰山总面积 426 平方千米,主峰玉皇顶海拔 1 532.8 米,山势雄伟壮丽,气势磅礴,名胜古迹众多,有"五岳独尊"之誉。孔子有"登泰山而小天下"之语;唐大诗人杜甫有"会当凌绝顶,一览众山小"的佳句。泰山在人们的心目中,已成为伟大、崇高的象征。

(http://www.china.com.cn/culture/txt/2006-12/13/content_7499466.htm)

这是网络上对山东泰山的一段介绍。辞藻华丽,还引用了唐代诗人的赞美诗句。

1. 词汇层

岳(高大的山) high mountain/sacred mountain;泰山 Mount Tai;衡山/恒山 Mount Heng;总面积 total area;主峰 highest peak;雄伟壮丽 imposing form;气势磅礴(磅礴:广大无边的) boundless grandeur;名胜古迹 scenic spots and historical sites;五岳独尊"Wu Yue Du Zun"(the most revered/esteemed of China's five sacred mountains)。

2. 句子层

E.g. 泰山古称岱山,又称岱宗。位于山东省中部,为中国五岳(泰山、华山、衡山、嵩山、恒山)之一。

英语是化零为整的语言,往往将汉语零碎、较短的句子用有形的词法和语法手段拉长,这样更易于英语读者获取信息,所以很多漂亮的英文句子都比较长。基于此,汉译英时译者须尽量将译文英语句子拉长。

本例中汉语包含两个句子,若译为两句英语句子则显得太过松散,英语的可读性低。这个汉语句子包含 4 个主谓结构,分别是"泰山古称……""……又称……""位于……""为……之一"。好的英语是将一句中多个主谓结构中的 1-2 主谓结构译作英语译文的主句,让其他主谓结构设法附着在主句主谓结构之上。就本句而言,可将"古称岱山,又称岱宗"顺译出来,作主语"泰山"的后置定语,将"位于"至于译文英语的谓语位置。为更清晰表意,可将"为……之一"单独译为一句。

"山"可用"mountain"或"Mount"来表达,如"泰山"可译作"the Taishan Mountain"或"Mount Tai",为节约篇幅起见,笔者采用"Mount Tai"译法。一般情况下,并列成分——罗列即可,但是本句中"衡山"和"恒山"译为英语时都是"Mount Heng",如果直接罗列处理,英语读者会视作重复,所以要另辟蹊径。笔者采取增加这两座山所在省名的方法加以区分。

本句可译为:"Mount Tai, called Mount Dai, also Mount Daizong in ancient times, is located in central Shandong Province. It is one of China's five high mountains (namely, Mount Tai, Mount Hua, Mount Song, and 2 Mounts of Heng, one in Hunan Province, and the other in Shanxi Province)."

E. g. 泰山总面积 426 平方公里,主峰玉皇顶海拔 1 532.8 米,山势雄伟壮丽,气势磅礴,名胜古迹众多,有"五岳独尊"之誉。

本句汉语包含 5 个逗号和 1 个句号。翻译时,译者一定要深挖信息之间的逻辑关系,确定出译文英语的主句,尽量将句子拉长。

对于汉语中的特色文化词,可采用"拼音+解释"的方法译为英语,这样做也是我们文化自信的表现。故此,笔者将"五岳独尊"译为"'Wu Yue Du Zun'(the most revered/esteemed of China's five sacred mountains)"。"有……之誉"地道的英文表达法是"enjoy the laudatory title of..."。笔者认为,可将"泰山总面积 426 平方公里,主峰玉皇顶海拔 1 532.8 米,山势雄伟壮丽,气势磅礴,名胜古迹众多"理解为"有'五岳独尊'之誉"的原因,以"泰山有……之誉"为译文英语的主句进行翻译。因此,本句可译为:"With a total area of 426 square kilometers and highest peak (Yuhuangding) rising 1532.8 meters above the sea level, Mount Tai, owing to its imposing form, boundless grandeur, and numerous scenic spots and historical sites, enjoys the laudatory title of 'Wu Yue Du Zun' (the most revered/esteemed of China's five sacred mountains)."其中,原句中的"总面积……"和"主峰……"被处理为伴随状语,用"with"引出,这样整句英语译文层次分明,表意清晰,读来地道、顺畅。

E. g. "登泰山而小天下。""会当凌绝顶,一览众山小。"

直到今天,汉语诗词英译依然是有争议的话题,翻译诗句会产生"千人千面"的效果,译文质量参差不齐,有些甚至令人堪忧。笔者相信"翻译就是翻译意思"的说法,为了保证译文的可读性,主张用现代英语翻译古汉语(包括诗词),在理解原文意思的基础上,用英语重现原文信息。这两句可分别译为:"Ascending Mount Tai, one gets the feeling that the world below is suddenly (very) small."和"Only when you stand on the top of the Mount Tai, can you see others as small."

【参考译文】

Mount Tai, called Mount Dai, also MountDaizong in ancient times, is located in central Shandong Province. It is one of China's five high mountains (namely, Mount Tai, Mount Hua, Mount Song, and 2 Mounts of Heng, one in Hunan Province, and the other in Shanxi Province). "Dongyue" (East Sacred Mountain), is so-named because of its location in the eastern part of China. With a total area of 426 square kilometers and highest peak (Yuhuangding) rising 1,532.8 meters above the sea level, Mount Tai, owing to its imposing form, boundless grandeur, and numerous scenic spots and historical sites, enjoys the laudatory title of "Wu Yue Du Zun" (the most revered/esteemed of China's five sacred mountains). Confucius said in his poem, "Ascending Mount Tai, one gets the feeling that the world below is suddenly (very) small; and Du Fu, a great poet of the Tang Dynasty, wrote the following line in one of his poems, "Only when you stand on the top of the Mount Tai, can you see others as small." Mount Tai has now become a symbol of greatness and nobility (majesty) in people's minds.

【例 2】

天安门广场

简介

　　天安门广场,位于北京市中心,地处北京市东城区东长安街,北起天安门,南至正阳门,东起中国国家博物馆,西至人民大会堂,南北长 880 米,东西宽 500 米,面积达 44 万平方米,可容纳 100 万人举行盛大集会,是世界上最大的城市广场。

　　广场地面全部由经过特殊工艺技术处理的浅色花岗岩条石铺成,中央矗立着人民英雄纪念碑和庄严肃穆的毛主席纪念堂,天安门两边是劳动人民文化宫和中山公园,与天安门浑然一体,共同构成天安门广场。1986 年,天安门广场被评为"北京十六景"之一,景观名"天安丽日"。

　　天安门广场记载了中国人民不屈不挠的革命精神和大无畏的英雄气概,五四运动、一·二九运动、五·二〇运动都在这里为中国现代革命史留下了浓重的色彩,同时还是无数重大政治、历史事件的发生地,是中国从衰落到崛起的历史见证。

　　2018 年 1 月 1 日 7 时 36 分,天安门广场升国旗仪式,首次由人民解放军仪仗队和军乐团执行。

建筑布局

　　明清时期,天安门广场是北京紫禁城正门外的一个宫廷广场,东、西、南三面用围墙围成一片普通百姓不可进入的禁地。广场北起天安门,南至正阳门,东起历史博物馆,西至人民大会堂,南北长 880 米,东西宽 500 米,面积达 44 万平方米。广场中央矗立着人民英雄纪念碑和庄严肃穆的毛主席纪念堂,广场西侧是人民大会堂,东侧面是中国国家博物馆,南侧是两座建于 14 世纪的古代城楼——正阳门和前门箭楼。天安门城楼坐落在广场的北端。天安门两边是劳动人民文化宫和中山公园,这些雄伟的建筑与天安门浑然一体构成了天安门广场,成为北京的一大胜景。

历史意义

　　天安门广场,北京市中心地带,中国政治活动中心,是世界上最大的城市中心广场。天安门广场不仅见证了中国人民一次次要民主、争自由,反抗外国侵略和反动统治的斗争,更是共和国举行重大庆典、盛大集会和外事迎宾的神圣重地,这里是中国最重要的活动举办地和和集会场所。……

(https://baike.so.com/doc/1519532-1606488.html)

　　这篇节选的网文简介了天安门广场的位置、规模、构成、建筑布局及其历史意义。

　　1. 词汇层

　　盛大集会 a grand rally;花岗岩 granite;人民英雄纪念碑 the Monument to the People's Heroes;毛主席纪念堂 Chairman Mao Memorial Hall;劳动人民文化宫 The Working People's Cultural Palace;浑然一体 integral part;不屈不挠的革命精神 unbending and unyielding revolutionary spirit;从衰落到崛起 from decline to resurgence;升国旗仪式 flag raising ceremony;仪仗队 Guard of Honor;军乐团 Military Band;紫禁城 the Forbidden City;人民大会堂 the Great Hall of the People;中国国家博物馆 the National Museum of China;外国侵略 foreign aggression;反动统治 reactionary rule。

　　2. 句子层

　　E. g. 天安门广场,位于北京市中心,地处北京市东城区东长安街,北起天安门,南至正阳门,东起中国国家博物馆,西至人民大会堂,南北长 880 米,东西宽 500 米,面积达 44 万平方米,可容纳 100 万人举行盛大集会,是世界上最大的城市广场。

　　本句一逗到底,共用了 11 个逗号,1 个句号。翻译时,第一印象是要断句,以意群为单位将这个句子断为 2~3 句英语句。

　　汉语的"动态"性质(多用动词表意)很明显,如"位于……""地处……""北起……""南至……""东起……""西至……""达……""容纳……""是……"。汉译英时,一定要设法将汉语的"动态"性质转化为英语的"静态"性质。笔者从"南北长 880 米"前断开,前后各一句。翻译时,要有很强的译入语词法、词法和语法意识,尽可能写出多样化的句子来。本句可译为:"Tian'anmen Square is located in the center of Beijing on East Chang'an Street in the Dongcheng District and stretches from Tian'anmen in the north to Zhengyang Gate in the south and from the National Museum of China in the east to the Great Hall of the People in the west. Measuring 880 meters from south to east and 500 meters from east to west, it is the world's largest urban square with an area of 440,000 m², which is spacious enough to accommodate 1 million people at a grand rally."

　　E. g. 广场地面全部由经过特殊工艺技术处理的浅色花岗岩条石铺成,中央矗立着人民英雄纪念碑和庄严肃穆的毛主席纪念堂,天安门两边是劳动人民文化宫和中山公园,与天安门浑然一体,共同构成天安门广场。1986 年,天安门广场被评为"北京十六景"之一,景观名"天安丽日"。

　　这里第一句汉语较长,陈述了天安门广场地面铺成材料和广场建筑构成 2 个概念。因此,翻译时先将本句断为 2~3 句。合理的断句其实是"化难为易",翻译工作会变得简单起来。第一句末尾信息与下一句主语同为"天安门广场",这是将第二句译为定语从句的典型情形。这样做后,译文英语一下子就拉长了。英语长句因明示的逻辑关系有时更易于读者获取信息。本句可译为:"The ground is paved with specially processed light-colored granite stones. In the center stand the Monument to the People's Heroes and the solemn Chairman Mao Memorial Hall; The Working People's Cultural Palace is on one side of the square and Zhongshan Park on the other. Both are integral parts of Tiananmen Square, which was rated in 1986 as one of Beijing's "16 beautiful landscapes and given a new name:'Bright Sunshine of Tian'anmen'."

E. g. 明清时期,天安门广场是北京紫禁城正门外的一个宫廷广场,东、西、南三面用围墙围成一片普通百姓不可进入的禁地。

  本句的核心意思是"天安门广场是宫廷广场,普通百姓不可进入。"翻译的困难在于译文英语句式的搭建上。笔者以"天安门广场是宫廷广场"作译入语的主句,将其他信息放置到它们应该出现的地方,然后将它们完形即可。本句可译为:"During the Ming and Qing Dynasties, this square, located outside the main gate of Beijing's Forbidden City and having walls on the eastern, western and southern sides — making it inaccessible to the common people — served as an imperial square."

E. g. 广场北起天安门,南至正阳门,东起历史博物馆,西至人民大会堂,南北长 880 米,东西宽 500 米,面积达 44 万平方米。

  这句所在段落信息部分与"简介"中的信息重复出现。为了保持网络信息的完整性,笔者在此加以保留。本句汉语的英译困难在于译文英语信息顺序的安排和逻辑关系的明晰上。虽然本句表达了"广场"的范围、长度、面积等概念,但是根据信息之间的紧密关系,笔者用"范围"信息搭建起了译文英语的句式,本句可译为:"The square, measuring 880 meters from south to east and 500 meters from east to west for an area of 440,000 m^2, stretches from Tian'anmen in the north to the Zhengyang Gate in the south and from the National Museum of China in the east to the Great Hall of the People in the west."

E. g. 天安门广场不仅见证了中国人民一次次要民主、争自由,反抗外国侵略和反动统治的斗争,更是共和国举行重大庆典、盛大集会和外事迎宾的神圣重地,这里是中国最重要的活动举办地和和集会场所。

  本句中有"不仅……更是……"关联词,可译为"not only... but also...";要注意细节之处的翻译。比如,"争自由"与其后的"反抗外国侵略"和"反动统治"不仅仅是表面上的并列关系,而是"非此即彼"的关系,即有"自由"就不能有"外国侵略"和"反动统治",前者"自由"和后两者是"反抗"(译作"against")的关系;翻译"共和国举行重大庆典、盛大集会和外事迎宾的神圣重地"时要合理安排信息顺序,正确措辞,然后一一罗列即可。本句可译为:"Tian'anmen Square not only has witnessed the Chinese people's struggles for democracy and freedom against foreign aggression and reactionary rule, but also serves as hallowed ground for China to hold major celebrations and grand rallies as well as to host foreign guests. It is here that China's most important activities and gatherings are held."

【参考译文】

Tian'anmen Square

Introduction

Tian'anmen Square is located in the center of Beijing on East Chang'an Street in the Dongcheng District and stretches from Tian'anmen in the north to Zhengyang Gate in the south and from the National Museum of China in the east to the Great Hall of the People in the west. Measuring 880 meters from south to east and 500 meters from east to west, it is the world's largest urban square with an area of 440,000 m^2, which is spacious enough to accommodate 1 million people at a grand rally.

The ground is paved with specially processed light-colored granite stones. In the center stand the Monument to the People's Heroes and the solemn Chairman Mao Memorial Hall; The Working People's Cultural Palace is on one side of the square and Zhongshan Park on the other. Both are integral parts of Tiananmen Square, which was rated in 1986 as one of Beijing's 16 beautiful landscapes and given a new name: "right Sunshine of Tian'anmen".

Being a witness to numerous political and historical events from China's decline to its resurgence, Tian'anmen Square embodies the Chinese people's unbending and unyielding revolutionary spirit and fearless heroism. It is also here that countless major political and historic events like the May 4th Movement, December 9th Movement and May 20th Movement have left their imprint on the history of modern Chinese revolution.

At 7:36 on January 1, 2018, the national flag raising ceremony was performed by the Chinese People's Liberation Army Guard of Honor and Military Band for the first time.

Architectural Layout

During the Ming and Qing Dynasties, this square, located outside the main gate of Beijing's Forbidden City and having walls on the eastern, western and southern sides — making it inaccessible to the common people — served as an imperial square. The square, measuring 880 meters from south to east and 500 meters from east to west for an area of 440,000 m^2, stretches from Tian'anmen in the north to the Zhengyang Gate in the south and from the National Museum of China in the east to the Great Hall of the People in the west. The Monument to the People's Heroes and the solemn (sombre) Chairman Mao Memorial Hall stand in the center; the Great Hall of the People is on the west side and the National Museum of China on the east side. The Zhengyang Gate and the Qianmen Embrasured Watchtower, which were built in the 14th Century, are on the south side. The Tiananmen Gatetower is located on the north end of the Tiananmen Square; The Working People's Cultural Palace is on one side of the square and Zhongshan Park on the other. These majestic structures together with the Tiananmen Gate constitute a single entity, Tiananmen Square, a major tourist attraction in Beijing.

Historical Significance

Located at the center of Beijing, Tian'anmen Square is the center of China's political activity and is the world's largest urban central square. Tian'anmen Square not only has witnessed the Chinese people's struggles for democracy and freedom against foreign aggression and reactionary rule, but also serves as hallowed ground for China to hold major celebrations and grand rallies as well as to host foreign guests. It is here that China's most important activities and gatherings are held...

练 习 题

一、请将以下文本译为汉语。

The city guide to York

It's time for the Romans and Vikings to make way for a new insurgency in the North Yorkshire jewel: a sparky, creative scenefuelled by innovative music, food and drink outlets.

What images come to mind when you think of York? The Minster, steam engines, Romans and Vikings, a city resisting the 21st century? But look beyond that twee facade, outside York's narrow medieval streets, and a very different city is asserting itself.

"It's definitely getting more vibrant," says Danielle Barge, editor of webzine *Arts York*. "In recent years, a lot of people have started independent projects: small theatre and film companies, artists' studios, music promoters. People are almost in artistic rebellion. They're taking it upon themselves to say, 'if no one else is going to make it, we will'."

Drop in at Earworm Records, advises local DJ and promoter Tor Petersen and, within minutes, you can plug yourself into a discerning local music scene — look out for events from Ouroboros; Young Thugs; Indigo_303; Animaux; Please Please You — that most visitors to York remain oblivious to. "It's a small city but there's a knowledgeable local crowd, and a very co-operative scene. We look out for each other. We struggle with a lack of venues and those we have, we really appreciate," says Petersen, who co-promotes the electronic music event series, Bad Chapel.

From the emerging, floating, grassroots Arts Barge Project to the imminent Spark York — a food and arts shipping container development with a distinctive community outreach angle — it is as if, after years of coasting on its historic attractions, York is waking up to the idea that both locals and visitors now demand more from a city. It's the same with food. York may still be dominated by chains and tourist-trap venues, but the individuality and ambition of key local independents, such as Skosh and Le CochonAveugle, have put the city on the map. It now looks like somewhere where interesting food can find an audience. For instance, hotly tipped chef Luke Cockerill (@lukecockerill) is working on a new York project.

But there are still barriers, and independents (in initiatives such as Fossgate's Sunday street parties) tend to help one another in the face of relentless external pressures.

"I moved to York in 2001 and it was seven years before it started to get any kind of decent restaurant scene, and that scene is getting more traction," says Ben Thorpe, editor at *York on a Fork* (yorkonafork.com). "All the independents I speak to have the same complaints about high [business] rates and unfair competition from chains."...

(https://www.theguardian.com/travel/2017/nov/15/alt-city-guide-york-food-music-art-theatre-drink-bars)

二、请将以下文本译为英语。

芬兰推"世上第一个昆虫面包" 每条含 70 只蟋蟀

2017 – 11 – 25 08:49:20 来源：中国新闻网（北京）

据外媒报道,芬兰烘焙食品公司 Fazer 23 日推出号称"世上第一个以昆虫为材料的面包"在店铺贩卖。每条面包要价 3.99 欧元（约合人民币 31 元）,比一般面包贵约两倍。

据悉,这款面包将干燥蟋蟀磨成粉末并加进面粉中制作,每条约含有 70 只蟋蟀。

Fazer 公司称,这种面包比起普通的含有更多蛋白质,且虫类具有脂肪酸、钙质、铁质和维生素 B12,对人体绝对是好处多多,但并未多加说明材料来源。

一名来自赫尔辛基的学生莎拉品尝完后称:"我没吃出什么差别,这款面包的味道就和一般的一样。"

Fazer公司研发主管薛贝可夫表示:"蟋蟀面包是很棒的蛋白质来源,也能让消费者越来越熟悉以昆虫为材料的食品。"他们预计慢慢将面包推广至全部的47家分店。

(http://travel.163.com/17/1125/08/D42UP9AN00067VF3.html)

第14章 新闻语篇翻译

14.1 新闻语篇的特点

"新闻"为外来语,对应英语"news"一词,是指报刊、电台、互联网等媒体用以记录和传播信息的一种文体,"对公众具有知悉意义或指导价值的事实报道,具有传输信息的功能"(汪洪梅,2017)。国内互联网各大网站充斥着对各行各业的新闻报道。新闻也许是最常见、公众最熟知的一种文体。

新闻报道的广泛性决定了新闻语篇内容的多学科性,如政治学、经济学、新闻学、传播学、社会学等,无论对新闻语篇写作还是翻译而言都提出了不小的挑战。

14.1.1 英语新闻语篇的特点

1. 标题特点

言简意赅,常有省略现象(以虚词为主,如冠词、连词"and"、人称关系代词、系动词"to be"等)。

E. g. President Xi Jinping's Special Envoy Wang Dongming to Attend the Inauguration Ceremony of President of Mauritania ("to attend..."前省略了"is")
习近平主席特使王东明将出席毛里塔尼亚总统就职典礼

E. g. Bulgarian President Rumen Radev to Visit China ("to visit..."前省略了"is")
保加利亚总统鲁曼·拉德夫将访华

E. g. The 11th Round of China-European Union (EU) Consultation on African Affairs Held in Beijing ("held"前省略了"was")
第11轮中欧非洲事务磋商在北京举行

E. g. Western partnership with Russia vital for peace ("vital"前省略了"is")
西方与俄罗斯的伙伴关系对和平至关重要

E. g. Sino-US relations still considered most important ("still"前省略了"is")
中美关系依然非常重要

如同一般文章标题一样,新闻标题一般不以句子面目出现(因为有省略现象),以突出核心信息。但是,从我国政府官网、权威报刊和杂志的英文报道来看,近年来英语句子形式的标题越来越普遍。

E. g. The 27th Round of Negotiation on Regional Comprehensive Economic Partnership

Kicks off in Zhengzhou

《区域全面经济伙伴关系协定》第二十七轮谈判在郑州开幕

E.g. Chinese and US Chief Trade Negotiators for High-level Trade Consultation Hold Telephone Conversation

中美经贸高级别磋商双方牵头人通话

E.g. China and Bangladesh Establish New Investment Cooperation Mechanism

中国与孟加拉国建立投资合作新机制

E.g. Zhong Shan Meets with Japanese Minister of Economy, Trade and Industry Seko Hiroshige

钟山会见日本经济产业大臣世耕弘成

E.g. Minister Zhong Shan Attends the Shanghai City Service Guarantee Work Promotion Conference of the 100-day Countdown of the 2nd China International Import Expo and Delivers a Speech

钟山部长出席第二届中国国际进口博览会倒计时100天上海城市服务保障工作推进大会并讲话

2. 词汇特点

(1) 专有名词多(因为新闻报道涉及人名、地名、机构名等)。

E.g. "Finally, we are in a position to start serious studies in this field after 10 years of preparation," said **Hiromitsu Nakauchi**, a researcher at **the Institute of Medical Science of the University of Tokyo**, according to *Asahi*. (粗体字为专有名词)

据《**朝日新闻**》报道,**东京大学医学研究所**的研究员**中内启光**说:"经过10年的准备,我们终于可以开始在这一领域进行严肃的研究。"

E.g. Chinese animators have been drawing inspiration from Chinese mythology ever since the country's first animated film, *Princess Iron Fan* in 1941, and in recent years its animated-film industry has released several hit films based on classic Chinese tales, such as *Monkey King: Hero is Back* in 2015, *Big Fish & Begonia* in 2016, and *White Snake* in 2019. (粗体字为专有名词)

自1941年中国第一部动画片《铁扇公主》问世后,中国的动画家们一直从中国的神话故事中汲取灵感。近年来,中国的动画电影产业推出了几部根据中国经典故事改编的热门电影,如2015年的《西游记之大圣归来》、2016年的《大鱼海棠》和2019年的《白蛇:缘起》。

(2) 用词往往简明扼要。集中表现是:用短词代替长词,词汇动态感强,表意生动形象。比如以下替代例子(用"="前词语代替"="后表达法):

cut = reduce/reduction; probe = investigate; bid = effort/attempt; deal = bargain/transaction; consensus = consensus of opinion; hurry = walk hurriedly; destroyed = totally destroyed; ban = prohibit;等等。

E.g. All this has happened not in spite of our plan to **cut** the deficit, but because of it.

所有这一切的发生并非与我们**削减**赤字计划无关,而是因削减赤字计划而产生。

E. g. The company will cooperate with any government **probe** in the matter as well.

该公司还将全力配合任何政府部门对该事件的**调查**。

E. g. Other countries appropriated the profits from the **deal**.

其他国家从这项**交易**中侵吞了利益。

E. g. Japan had previously **banned** such experiments.

日本此前曾**禁止**此类实验。

(3)新闻报道中缩略语多,主要涉及国家名、机构名、专业术语等方面。

E. g. "That includes very much that they will one day be members, if they so wish of course, and important to add, when they meet **NATO** standards," he said.

他说:"如果这些国家达到**北大西洋公约组织**标准,他们也希望成为北约成员,未来某一天当然可以将他们加入到北约名单中,成为会员。这是极有可能发生的事情。"

E. g. You have quite an imagination! I wish you were working for **FBI**.

你还真有想象力。真希望你是在**联邦调查局**工作的。

E. g. Inside every cell in all organisms, there are strands of **DNA**.

所有生物的每个细胞内部都存在 DNA 分子链。(DNA＝ deoxyribonucleic acid 脱氧核糖核酸)

14.1.2 汉语新闻语篇的特点

(1)新闻内容具体、准确。新闻要求具体、如实报道事件涉及的人、发生的时间、地点、经过、结果。

(2)报道的客观性。新闻报道须客观,不夸大也不缩小,不因报道者的民族身份、宗教信仰、政治立场等因素而使其客观性发生改变。

(3)语言风格须简洁明了,通俗易懂。由于其时效性,新闻语言要开门见山、直截了当、简明扼要,不拖泥带水。新闻语言要通俗易懂,明白晓畅,易于普通大众获取信息。

14.2 新闻语篇翻译实例

14.2.1 英译汉

China urges all parties to preserve Iran nuclear deal

China hopes all relevant parties will maintain their commitment to the Iranian nuclear deal, China's Ministry of Foreign Affairs said Friday.

"We believe this deal is important to ensuring the international nuclear nonproliferation regime and regional peace and stability. "We hope all parties can continue to preserve and implement this deal, ministry spokesperson Hua Chunying said at a regular press briefing.

China's top diplomat, Yang Jiechi, discussed the Iranian nuclear issue with US Secretary of State Rex Tillerson in a phone call on Thursday to prepare for President Donald Trump's November visit to Beijing, Hua said.

The Iranian nuclear agreement was signed between Iran and six world powers — Britain, China, France, Germany, Russia and the US — at talks coordinated by the EU.

While the deal stalled Iran's nuclear program and thawed relations between Tehran and its "Great Satan", opponents say it also prevented efforts to challenge Iranian influence in the Middle East.

US officials say Trump will not kill the international accord outright, instead "decertifying" the agreement and leaving US lawmakers to decide its fate.

UN nuclear inspectors have said that Iran is meeting the technical requirements of its side of the bargain, dramatically curtailing its nuclear program in exchange for sanctions relief.

Iran President Hassan Rouhani lashed out at his US counterpart saying he was opposing "the whole world" by trying to abandon the agreement.

Russian Foreign Minister Sergei Lavrov on Friday told his Iranian counterpart Mohammad Javad Zarif that Russia remains fully committed to the deal, Russia's foreign ministry said in a statement.

Lavrov told Zarif in a telephone conversation that Moscow was firmly determined to implement the deal in the form in which it was approved by the UN Security Council, the ministry said.

If the US leaves the Iran nuclear deal, this will be the end of this international agreement, the TASS news agency cited Iranian parliament speaker Ali Larijani as saying.

Larijani, in Russia's second largest city of St. Petersburg for an international parliamentary forum, also said that Washington's withdrawal from the nuclear deal could lead to global chaos, TASS reported.

Iran hopes that Russia will play a role in resolving the situation around the nuclear deal, Larijani said meeting Vyacheslav Volodin, the speaker of the Duma lower house of Russia's parliament, the Interfax news agency reported.

(Global Times, October 14 – 15, 2017)

通过这篇新闻报道,我们了解到美国退出伊朗核协议引发联合国、欧盟、伊朗、俄罗斯和中国等机构和国家的反应。中国敦促各方遵守伊朗核协议。

1. 词汇层

Iran nuclear deal 伊核协议;China's Ministry of Foreign Affairs 中国外交部;the international nuclear nonproliferation regime 国际核不扩散机制;Secretary of State 国务卿;world power 世界大国;stall 停止;thaw 缓解;decertify 收回……的证件;curtail 削减;lash out at 猛烈抨击;Hassan Rouhani 哈桑·鲁哈尼;Sergei Lavrov 谢尔盖·拉夫罗夫;Mohammad Javad Zarif 穆罕默德·贾瓦德·扎里夫;the TASS news agency 塔斯通讯社;Ali Larijani 阿里·拉里贾尼;St. Petersburg 圣彼得堡;VyacheslavVolodin 维亚切斯拉夫·沃洛金;the Interfax news agency 国际文传通讯社。

2. 句子层

E. g. China hopes all relevant parties will maintain their commitment to the Iranian nuclear deal, China's Ministry of Foreign Affairs said Friday.

本句为倒装句，句子真正的主语为"China's Ministry of Foreign Affairs"。这是典型的新闻报道模式：采用倒装句式，将最重要信息置于句首，加以突出（本文中还有其他句子亦如此）。翻译为汉语时，要按照汉语的表述习惯（什么时间，谁说了什么话）将语序加以调整。因此，本句可译为："周五，中国外交部发表声明称，中国希望相关各方恪守对伊核协议的承诺。"

E. g. China's top diplomat, Yang Jiechi, discussed the Iranian nuclear issue with US Secretary of State Rex Tillerson in a phone call on Thursday to prepare for President Donald Trump's November visit to Beijing, Hua said.

本句为倒装句，以突出讲话内容，主句为"Hua said"。笔者认为，若将本句中的"top diplomat"直译为"高级外交官"，一定程度上破坏了表达的严肃性，不如还原为身份"外交部长"。"to prepare..."为目的状语，顺译即可。本句可译为："华春莹说，周四中国外交部长杨洁篪与美国国务卿雷克斯·蒂勒森电话讨论了伊核问题，为美国总统唐纳德·特朗普11月份访问北京做好准备。"

E. g. While the deal stalled Iran's nuclear program and thawed relations between Tehran and its "Great Satan", opponents say it also prevented efforts to challenge Iranian influence in the Middle East.

本句结构是："状语＋主句（主语＋谓语＋宾语从句）"，意思不难理解。翻译时要稍加调整，以"反对者声称"（"opponents say"）作译文汉语的主句，表达起来会顺畅很多。本句可译为："反对者声称，尽管伊核协议使伊朗停止其核计划，缓解了德黑兰与'大撒旦'之间的敌对关系，但同时也阻碍了为挑战伊朗在中东影响力而做出的努力。"

E. g. UN nuclear inspectors have said that Iran is meeting the technical requirements of its side of the bargain, dramatically curtailing its nuclear program in exchange for sanctions relief.

英语新闻报道中的"say"不要总是译为"说"，更多的时候要译为"表示""声称"等意思。原文英语句中的"dramatically curtailing..."作状语，而汉语是"竹节式"语言（主谓结构按逻辑关系顺次罗列），依次顺译即可。此外，"in exchange for"是静态英语的表达法，翻译时要转换为动态的汉语"以换取"。本句可译为："联合国核查人员表示，伊朗正在逐步满足谈判确定的伊方技术要求，大幅削减核计划，以换取国际社会取消制裁。"

E. g. Russian Foreign Minister Sergei Lavrov on Friday told his Iranian counterpart Mohammad Javad Zarif that Russia remains fully committed to the deal, Russia's foreign ministry said in a statement.

这是本篇新闻稿中又一句倒装句，翻译时要注意调整信息顺序。因为主语是"Russia's foreign ministry"，"said"不宜译为"说"（主语是人时可以这样翻译），也不要译为"表示"（因为实际上并没有表达观点），而要译为"提到"/"谈到"，因为放到句首的核心信息是事件本身。因此，本句可译为："在一份声明中，俄罗斯外交部提到，周五俄罗斯外交部长谢尔盖·

拉夫罗夫告诉伊朗外交部长穆罕默德·贾瓦德·扎里夫，俄罗斯仍将继续执行该协议。"

【参考译文】

中国敦促各方遵守伊核协议

周五，中国外交部发表声明表示，中国希望相关各方恪守对伊核协议的承诺。

中国外交部发言人华春莹在例行记者会上表示："我们认为伊核协议对于确保国际核不扩散机制、促进中东地区和平稳定发挥了重要作用。我们希望各方继续遵守并执行这一协议。"

华春莹说，周四中国外交部长杨洁篪与美国国务卿雷克斯·蒂勒森电话讨论了伊朗核问题，为美国总统唐纳德·特朗普11月份访问北京做好准备。

在欧盟的积极斡旋下，伊朗与世界六大国（英国、中国、法国、德国、俄罗斯和美国）签订了该协议。

反对者声称，尽管伊核协议使伊朗停止其核计划，缓解了德黑兰与"大撒旦"之间的敌对关系，但同时也阻碍了为挑战伊朗在中东影响力而做出的努力。

美国官员称，特朗普总统不会直接废除该国际协议，而是收回该协议签署文件，由美国国会决定其最终命运。

联合国核查人员表示，伊朗正在逐步满足谈判确定的伊方技术要求，大幅削减核计划，以换取国际社会取消制裁。

伊朗总统哈桑·鲁哈尼抨击美国总统时声称，特朗普试图通过放弃该协议与"全世界"为敌。

在一份声明中，俄罗斯外交部提到，周五俄罗斯外交部长谢尔盖·拉夫罗夫告诉伊朗外交部长穆罕默德·贾瓦德·扎里夫，俄罗斯仍将继续执行该协议。

俄罗斯外交部声称，拉夫罗夫在电话交谈中告诉扎里夫，俄罗斯政府将继续按照联合国安理会认可的方式执行伊朗核协议。

塔斯通讯社援引伊朗议会议长阿里·拉里贾尼的声明称，如果美国退出伊核协议，即表明该国际协议终止执行。

据塔斯通讯社报道，在俄罗斯第二大城市圣彼得堡召开的国际议会论坛上，拉里贾尼表示，美国政府退出伊核协议可能会引发全球混乱。

国际文传通讯社消息称，拉里贾尼在会见俄罗斯国家杜马下议院议长维亚切斯拉夫·沃洛金时指出，在解决有关伊核协议问题上伊朗呼吁俄罗斯发挥重要作用。

（《环球时报》，2017年10月14日—15日）

14.2.2 汉译英

【例1】

2018年8月17日，马来西亚总理马哈蒂尔应邀抵达中国，开始为期5天的访华之旅。据悉，马哈蒂尔首先来到杭州，到访阿里巴巴总部，并与阿里巴巴董事局主席马云举行会面。杭州是马哈蒂尔此次中国行的第一站，也是92岁的马哈蒂尔二度出任总理后，首次出访中国。中国成为其就任以来在东盟之外首个正式出访的国家。据悉，在杭州，马哈蒂尔还将访问吉利汽车，之后将前往北京。

(http://mini.eastday.com/a/180818101758762.html)

在马哈蒂尔第二次当选马来西亚总理后,这是环球网对其行程做的简短报道。

1. 词汇层

马来西亚总理马哈蒂尔 Malaysian Prime Minister Mahathir Mohamad;应邀 at the invitation of;阿里巴巴总部 the headquarters of Alibaba;董事局主席 chairman of the board directors;东盟 ASEAN（Association of Southeast Asian Nations）;吉利汽车 Geely Automobile。

2. 句子层

笔译是严肃的目的语写作,新闻语篇翻译也不例外。译者须居高临下,高瞻远瞩,在准确理解原文的基础上,尽量用最准确的措辞、最凝练的语言来表达最清晰的概念。

E.g. 2018年8月17日,马来西亚总理马哈蒂尔应邀抵达中国,开始为期5天的访华之旅。

新闻有很强的时效性,时间概念很重要,中外相同。汉语新闻写作时,按照"状语先行"的原则,一般放到句首;译为英语语篇时,最好也放到句首,顺译即可,这也是翻译时节省精力的做法。

"应邀抵达中国"不可译为"was invited to arrive in China",这不是地道的英文表达方式。我们可以说"be invited to deliver a speech"或"be invited to address students at a ceremony"等,但不可说"be invited to arrive in somewhere",逻辑不通。原文涉及1个人称:"马来西亚总理马哈蒂尔"（主语）,发出3个动作:"应邀"、"抵达"和"开始"（谓语）。好的英语句子表达往往以一个谓语动作为中心,与主语形成译文的主句结构,其他谓语设法附着在主句的主谓结构上,这样写出的句子有层次感。这是英语句子表达的本质要求,也是英汉差异转换的结果,切不可像汉语表达那样,将一个个动作罗列而已,导致英语译文可读性低。本句可译为:"On August 17, 2018, the 92-year-old Malaysian Prime Minister Mahathir Mohamad arrived in China as an invited guest for a five-day tour of the country."为信息表达更紧凑起见,笔者将下文中的"92岁"译在了这里。

E.g. 据悉,马哈蒂尔首先来到杭州,到访阿里巴巴总部,并与阿里巴巴董事局主席马云举行会面。杭州是马哈蒂尔此次中国行的第一站,……

"据悉"译为模块化的"It is reported..."（因为这是新闻报道,此处最好不用"It is said..."或者"It is known..."等译法）。同样是为了把信息表达得紧凑些,下一句的"杭州是马哈蒂尔此次中国行的第一站"可提到本句,放到"来到杭州"后,译作"his first stop",作

"Hangzhou"的同位语。本句可译为:"It is reported that he first visited the headquarters of Alibaba in Hangzhou, his first stop, and met with Ma Yun (Jack Ma), chairman of the board directors."

【参考译文】

On August 17, 2018, the 92-year-old Malaysian Prime Minister Mahathir Mohamad arrived in China as an invited guest for a five-day tour of the country. It is reported that he first visited the headquarters of Alibaba in Hangzhou, his first stop, and met with Ma Yun (Jack Ma), chairman of the board directors. This is Mahathir's first official visit to a non-ASEAN country since becoming prime minister for the second time. In Hangzhou, he will also visit Geely Automobile before heading to Beijing.

【例 2】

美芝加哥"血腥"圣诞周末　12 人被枪杀　40 多人受伤

中新社休斯敦 12 月 26 日电　美国芝加哥市在刚刚过去的圣诞节周末再次贴上了"血腥"标签。警方称,自 23 日晚至 26 日清晨,全市共有 12 人被枪杀,40 多人遭枪击受伤,暴力案件比去年同期大幅增加。

"如果任何人认为这是正常的,那么你错了,这不可以,这种现象非常不正常。"芝加哥警方负责人约翰逊(Eddie Johnson)在 26 日的新闻发布会上表示,这种暴力是"不可接受的"。去年圣诞节周末,芝加哥有 30 多人被枪击,6 人死亡。

据《芝加哥太阳时报》和美国广播公司报道,警方指出,90%的死者与帮派有关系,他们有犯罪记录或已被列在警局执法列表上。大多数人都是被有针对性地枪杀。

约翰逊再次呼吁对重复犯罪的罪犯进行更严厉的处罚,称自周五以来,警方已从街上收缴了 45 支枪。"犯罪分子的目标非常明确,专门在那些潜在竞争对手与家人朋友庆祝节日时下手。"约翰逊说,"这些枪手罪犯已经多次向我们表明,他们不会按社会规则行事。"

最近的一次大规模枪击案发生在 25 日晚 9 时 20 分,芝加哥东查塔姆社区一家举行聚会时,一人身穿灰色兜帽衫从相邻小巷走出并朝门廊开火,2 名青年遭枪击身亡,另有 5 人受伤,其中两人情况危急。

芝加哥警察局发言人埃斯特拉达(Jose Estrada)告诉《纽约时报》,今年迄今该城市死于凶杀案的人数达 745 人,高于去年同期的 476 人,是近 20 年来的最高纪录。遭枪击的受害者总数为 4 252 人,比 2015 年的 2 884 人增加了 47%。大多数受害者为 30 岁以下男性。

当局将凶杀和枪击案数量暴增归因为帮派拼斗,各帮派团伙常因领地等纠纷恶化升级为枪战。由于凶杀案增加速度太快,许多案件警方调查至今没有拘捕任何嫌犯。

今年,芝加哥谋杀案数量已经超过洛杉矶和纽约市的总和。《芝加哥论坛报》刊发专栏文章称,该市近 20 年来凶杀死亡人数首次超过 700 人,"罪恶之城"已全面失控。

(http://news.qq.com/a/20161227/013540.htm)

根据这篇新闻报道,我们得知圣诞周末美国"罪恶之城"芝加哥市发生枪击案件,造成 12 人被枪杀、40 多人遭枪击受伤的后果,引发警局担忧和民众心理恐慌。

1. 词汇层

"血腥"圣诞周末"Bloody" Christmas weekend;中新社休斯敦 the China News Service in Houston;暴力案件 violent incident;《芝加哥太阳时报》the Chicago Sun-Times;美国广播公司 ABC(American Broadcasting Company);帮派 gang;执法列表 law enforcement list;收缴 confiscate;兜帽衫 hooded shirt;帮派拼斗 battles between gangs;《芝加哥论坛报》Chicago Tribune。

2. 句子层

E.g. 中新社休斯敦12月26日电　美国芝加哥市在刚刚过去的圣诞节周末再次贴上了"血腥"标签。

"中新社休斯敦12月26日电"是中文新闻标准的写法,表明这条新闻是某家新闻机构几月几日报道的。翻译时,按照汉语意思译为英语即可,如将本句译为"As reported by the China News Service in Houston on Dec. 26,…",与其后发生的事件本身译为一句。本句可译为:"As reported by the China News Service in Houston on Dec. 26, Chicago was once again marked by "bloodiness" this past Christmas weekend."

E.g. 警方称,自23日晚至26日清晨,全市共有12人被枪杀,40多人遭枪击受伤,暴力案件比去年同期大幅增加。

"警方称"后跟了3个主谓结构:"共有12人被枪杀""40多人遭枪击受伤"和"暴力案件增加",安排好信息顺序,一一罗列即可。本句可译为:"The police stated that 12 persons were killed and over 40 injured in shootings which took place between the evening of the 23rd and the morning of the 26th, and that violent incidents had greatly increased as compared with the previous year."

E.g. 据《芝加哥太阳时报》和美国广播公司报道,警方指出,90%的死者与帮派有关系,他们有犯罪记录或已被列在警局执法列表上。

"据……报道"对应英语"reported by…"。"警方指出"后有两个主谓结构,并列罗列即可。本句可译为:"As reported by *the Chicago Sun-Times* and ABC, the police indicated that 90% of the deceased were involved with gangs and that they either had criminal records or had been entered on the law enforcement list at the police station."

E.g. 最近的一次大规模枪击案发生在25日晚9时20分,芝加哥东查塔姆社区一家举行聚会时,一人身穿灰色兜帽衫从相邻小巷走出并朝门廊开火,2名青年遭枪击身亡,另有5人受伤,其中两人情况危急。

本句的翻译成功与否取决于译者能否合理断句。根据意群,本句包含3个意思,即"最近的一次大规模枪击案发生在……","一家举行聚会……"和"开火"造成的后果"2名青年身亡""5人受伤"。为表意清晰起见,可译作3句,运用正确的词法、句法和语法一一翻译即可。本句可译为:"The most recent mass shooting occurred at 9:20 on the evening of the 25th. A family party was being held in East Chatham Community in Chicago, when a person wearing a grey hooded shirt stepped out of an alley and opened fire at the porch. 2 young people were shot dead and 5 were injured, 2 critically."

E.g. 芝加哥警察局发言人埃斯特拉达(Jose Estrada)告诉《纽约时报》,今年迄今该城市死于凶杀案的人数达745人,高于去年同期的476人,是近20年来的最高纪录。

"芝加哥警察局发言人埃斯特拉达"是汉语典型的表达方法:先交代身份,再给出人名,先铺陈,后引出主人翁。然而,英语往往先交代人名,然后视情况处理身份信息:若信息短,处理成同位语(如此处:"Jose Estrada, spokesman for the Chicago Police Department");若信息较长,可处理为一个英语句子(或从句)。本句"《纽约时报》"后信息的话题是"人数"("达……""高于……""是……"),用作英语宾语从句的主语,翻译困难即可迎刃而解。"是近20年来的最高纪录"无法与之前信息写在一起,断为另一句即可。本句可译为:"Jose Estrada, spokesman for the Chicago Police Department, told *the New York Times*, that the number of murder cases has reached 745 so far, an increase of 476 as compared with the same period last year. It has been record-breaking over the last 20 years."

E. g. 《芝加哥论坛报》刊发专栏文章称,该市近20年来凶杀死亡人数首次超过700人,"罪恶之城"已全面失控。

此处"称"对报纸而言,对应的英语是"report"(报道),其后有两个主谓结构,作宾语,并列起来,一一罗列即可。本句可译为:"*The Chicago Tribune* reported in the special column that the death toll exceeded 700 for the first time in 20 years, and that crime in the city has been totally out of control."

【参考译文】

"Bloody" Christmas Weekend in Chicago
Resulted in 12 Persons Shot Dead and Over 40 Persons Injured

As reported by the China News Service in Houston on Dec. 26, Chicago was once again marked by "bloodiness" this past Christmas weekend. The police stated that 12 persons were killed and over 40 injured in shootings which took place between the evening of the 23rd and the morning of the 26th, and that violent incidents had greatly increased as compared with the previous year.

Eddie Johnson, who is in charge of the Chicago police, said at a press conference on the 26th, "If anyone thinks this is normal, then you are wrong. Such phenomena (occurrences) are very abnormal". Such violence is "unacceptable". Last year on Christmas weekend, over 30 persons were shot, and 6 killed in Chicago.

As reported by *the Chicago Sun-Times* and ABC, the police indicated that 90% of the deceased were involved with gangs and that they either had criminal records or had been entered on the law enforcement list at the police station. Most had been targeted to be shot dead (killed).

Johnson called for more severe punishment of criminals who repeatedly commit crimes. He said that the police had confiscated 45 guns from the street since Friday. Johnson said, "The criminal has a very clear target and kills potential competitors, especially when they are celebrating an occasion with their family and friends". "These criminal gunmen have repeatedly demonstrated that they will not observe social rules".

The most recent mass shooting occurred at 9:20 on the evening of the 25th. A family party was being held in East Chatham Community in Chicago, when a person wearing a grey hooded shirt stepped out of an alley and opened fire at the porch. 2 young people were shot dead and 5 were injured, 2 critically.

Jose Estrada, spokesman for the Chicago Police Department, told *the New York Times*, that the number of murder cases has reached 745 so far, an increase of 476 as compared with the same period last year. It has been record-breaking over the last 20 years. The total number of victims is 4,252 this year, a 47% increase over 2015 when there were 2,884 victims. Most victims were men under the age of 30.

Authorities attribute the sharp increase in murders and shootings to battles between gangs. Territorial disputes often deteriorate into gunfights. Due to the rapid increase in murder cases, many suspects have, so far, not been arrested for police investigation.

This year, the number of murder cases in Chicago has exceeded the total number of Los Angeles and New York. *The Chicago Tribune* reported in the special column that the death toll exceeded 700 for the first time in 20 years, and that crime in the city has been totally out of control.

练 习 题

一、请将下列文本译为汉语。

Illegal Loggers Arrested & Fined for Unlawful Possession of Forest Produce
Belmopan: 28th September 2017

Pablo Valladarez, a 47-year-old resident of Guinea Grass Village, Orange Walk District, was fined \$22,838 for illegal possession of Mahogany yesterday. Police and Forest Department officials intercepted Valladarez, CatalinoCopo, Nelson Mosa, and Aurelio Mai, residents of Guinea Grass and Santa Martha villages, hauling illegally harvested Mahogany on the Old Northern Highway, Orange Walk District, on 24th September 2017.

The Forest Department confiscated 34 unstamped Mahogany logs, equivalent to 1,268.8 board feet, and charged the four men for "Unlawful Possession of Forest Produce" under *the Forest Act* of 2000 and its 2017 Amendment. On 27th September, the men appeared before Magistrate Albert Hoare in the Orange Walk Magistrate Court, whereValladarez pleaded guilty to the charges. As a result, Valladarez may face five years imprisonment if he is unable to pay the fine by 31st January 2018.

Prior to this incident, the Forest Department also intercepted Juan Reyes, a resident of Carmelita Village, Orange Walk District, in possession of multiple timber species such as Negrito, Santa Maria and Mahogany. He was consequently fined for Unlawful Possession of Forest Produce on 19th September 2017.

"Our Department is seriously cracking down on this pervasive problem of illegal logging," said Saul Cruz, Officer in Charge, Orange Walk Range, Forest Department. Cruz explained that timber harvesting can be highly lucrative but illegal logging destroys our opportunities. "It is a forest crime that threatens the future of our sustainable logging industry, our environment, the livelihoods of forest-dependent people, and our local and national economy," he added.

The passage of the 2017 Amendment to *the Forest Act* provides stricter penalties to match the seriousness of the offences but the Department recognized that laws alone are insufficient. As a result, the Department continues to step-up efforts by forging stronger ties with other law enforcement officials, building community awareness, and increasing monitoring and law enforcement exercises.

The department reminds the public that the logging season is closed from June 15th to October 15th and any logging during this period is strictly prohibited. This ensures safer logging operations, protect forest soils and reduce damage to public roads during the rainy season. Report forest crimes to the Forest Department at 822-2079 or email at info@forest.gov.bz.

(http://www.guardian.bz/index.php?option=com_content&view=article&id=14235:illegal-loggers-arrested-a-fined-for-unlawful-possession-of-forest-produce&catid=39:crime&Itemid=73)

二、请将下列文本译为英语。

习近平和夫人彭丽媛举行宴会欢迎特朗普夫妇

李克强、张德江、俞正声、张高丽、栗战书、汪洋、王沪宁、赵乐际、韩正、刘云山、王岐山等出席。

新华社北京11月9日电（记者谭晶晶、许可） 国家主席习近平和夫人彭丽媛9日晚在人民大会堂举行宴会，欢迎美国总统特朗普和夫人梅拉尼娅。李克强、张德江、俞正声、张高丽、栗战书、汪洋、王沪宁、赵乐际、韩正、刘云山、王岐山出席。

人民大会堂金色大厅金碧辉煌，高朋满座，两国元首在宴会上分别发表了热情洋溢的讲话。

习近平指出，中美两国虽然远隔重洋，但地理距离从未阻隔两个伟大国家彼此接近。双方寻求友好交往和互利合作的努力从未停息。自45年前尼克松总统访华、开启中美重新交往大门以来，在中美几代领导人和两国人民的共同努力下，中美关系实现了历史性发展，造福了两国人民，改变了世界格局。今天中美关系已经变成你中有我、我中有你的利益共同体。现在，两国在维护世界和平、促进共同发展方面拥有更多、更广的共同利益，肩负更大、更重的共同责任，中美关系的战略意义和全球影响进一步上升。

习近平强调，特朗普总统此次访华具有重要的历史意义。两天来，我同特朗普总统共同规划了未来一个时期中美关系发展的蓝图。我们一致认为，中美应该成为伙伴而不是对手，两国合作可以办成许多有利于两国和世界的大事。中国古人说，"志之所趋，无远勿届，穷山距海，不能限也"。我坚信，中美关系面临的挑战是有限的，发展的潜力是无限的。只要本着坚韧不拔、锲而不舍的精神，我们就一定能谱写中美关系新的历史篇章，中美两国一定能为人类美好未来做出新的贡献。

特朗普表示，感谢习近平主席热情接待他到访中国这个伟大的国度。他说："美国人民十分景仰中国悠久的文明传承。在此历史性时刻，他深信美中合作可以造福美中两国人民，并为世界带来和平、安全与繁荣。"

宴会开始前，金色大厅内回放了习近平主席同特朗普总统海湖庄园会晤、汉堡会晤以及特朗普总统此次访华的经典片段。特朗普总统并现场提议播放了他外孙女用中文演唱中国歌曲、背诵《三字经》和中国古诗的视频。现场一再响起热烈掌声。

出席宴会的还有丁薛祥、王晨、刘鹤、刘延东、杨洁篪、郭声琨、蔡奇、韩启德、董建华、万钢、周小川以及美国国务卿蒂勒森等多位内阁成员、白宫高级官员。（完）

责编：王民和

(http://news.ifeng.com/a/20171109/53124003_0.shtml)

第 15 章 医学语篇翻译

"医学是通过科学或技术手段处理人体各种疾病或病变的学科。"(360百科)。人类的发展过程也是与疾病不断抗争的过程。人类预期寿命大大延长,这与全世界医学科学家、医务工作者及相关人员的刻苦钻研、开拓进取和默默奉献关系紧密。这些人员至今已积累了大量的医学文献供更多人阅读学习。科研(包括医学领域)工作者在线发表论文(尤其是英语论文),分享研究成果,希望实现全世界共同进步。

在"一带一路"建设的大背景下,我国对外交流与合作日益频繁,社会对专业人才(包括医学领域)需求量越来越大。医药科研工作者、医药卫生行业的工作者、在校医学专业学生等人员不但要不断提高自己的专业水平,还要提高自己用外语对外交流能力。医学语篇翻译能力是"用外语对外交流能力"的一种表现,有利于医药从业者、医学专业学生学习更先进的医学技术,推动本国医学向前发展,也有利于向世界分享我国的医学成果,造福全人类。

为保证医学语篇翻译质量,译者须了解医学语篇的文体特点、医学术语和医学原理,具备扎实的双语能力、翻译专业知识,还须进行大量的医学语篇翻译实践。

15.1 医学语篇的特点

1. 术语丰富

医学分为基础医学和临床医学,可进一步分为法医学、检验医学、预防医学、保健医学、康复医学等(百度百科)。每一分支又可以进一步细分。时至今日,医学已大大发展了,不断向纵深发展,产生了大量的医学术语。这些术语大量、频繁地出现在医学专业文章中。

2. 表述客观

医学是性命攸关的学科。医药知识是不以人的意志为转移的科学存在,用文字表述时,来不得半点马虎。医药成分、药量、服用时间、服用频次、服用人群、服用禁忌等须表述客观,否则可能会造成严重的后果。

3. 措辞准确

"医学和有关学科的名词术语,包括单词、词组、缩写、前缀与后缀等有共约16万条"(百度百科),来源广泛,构词法多样,稍有疏忽可能就会造成表意不准确,甚至出现错误,所以医学语篇中措辞务必准确。

4. 语法规范

由于医学语篇表述的客观性要求,清晰的表意不但来自措辞的准确性,还要求语法的规范性。医学专业有些知识和原理错综复杂,表述时要求有更高的语法意识,要严格按照书面语言的语法规则表意,不可掉以轻心。

5.文体质朴

医学语篇属于"信息型文本",主要功能是提供医学信息,供读者学习或公司应用。因此,医学语篇文体质朴,文本中没有华丽的辞藻,也没有很多修辞手法的应用情形。

15.2 医学语篇翻译实例

15.2.1 英译汉

What Your Mouth Can Tell Your Dentist

[More than forty diseases-some of them life-threatening-can be detected during an oral examination.]

The pain in David Rogers' lower jaw was getting worse. It was beginning to interfere with his concentration. Reluctantly, he took time from work to visit the nearest dentist.

The dentist made a thorough examination. He even took Rogers' blood pressure. "Your heart beat sounds very irregular to me," he said as he removed the stethoscope. "It may be normal for you, but I think we should check it out before we go any further."

Rogers protested. Except for his aching lower jaw, he felt fine. But the dentist was insistent, so Rogers finally gave in and went to see a physician in the same building. "Anything to get rid of this pain in my mouth," he muttered.

The pain, it turned out was a lifesaver, for the physician discovered that Rogers was in the early stages of a heart attack. Fortunately, he was able to abort the attack before it did serious damage. Rogers was lucky. His dentist had known that lower-jaw pain may be a symptom of a heart attack. A few months later, the grateful Rogers returned to the dentist, this time to thank him. "You probably saved my life," he told him.

Lucy Parker rarely got cavities. Nevertheless, she believed in regular dental checkups.

During one checkup, her dentist found a whitish patch. It turned out to be cancer. In her case, there was no doubt. The checkup really did save her life.

Both David's and Lucy's stories are true. They illustrate the vital importance of regular trips to the dentist. A six-month checkup can provide the first warning not only of dental disease, but of general health problems, as well.

"Cancer and heart disease are the two biggies that kill people," says Dr. Michael Roberts, the chief of patient care at the National Institute of Dental Research. "But they are not the only health problems that can be spotted by a dentist. According to the American Dental Association, more than 40 serious ailments, including diabetes, bulimia, brain tumors, and AIDS, can be detected in the mouth."

"The mouth is the most visible and accessible barometer of the body's health," says

Dr. Lawrence Cohen, chairman of the Department of Dentistry at the Illinois Masonic Medical Center. "It's a mirror of disease, because it's such an easily observable area. All you need is a bright light and a dental mirror. You don't have to use tubes as you do to see into the gastrointestinal tract."

The American Cancer Society estimates that this year alone 29,000 patients will be diagnosed with oral cancer, and of this number, 9,500 will die. "If you go regularly to your dentist," says Dr. Diane Stern, a Florida oral pathologist, "the likelihood is that, if you develop oral cancer, it will be spotted in an early stage. And the smaller the malignancy is at the time of diagnosis, the more likely you are to be cured." Dr. Jerry Rosenbaum, Florida periodontist, agrees: "A patient's survival rate is directly related to the time the cancer is found."

Dentists may also detect numerous health problems that are not life threatening. "One of the most common conditions people get is canker sores," Dr. Cohen says. "If you deal with the younger age group, then you see a lot of the acute infections such as herpes and trench mouth, which is due to bacteria and causes open sores between the teeth. Other symptoms may be pain and a bad taste in the mouth."

Certain blood diseases, such as anemias, can also be detected in the mouth very early, Dr. Cohen adds: "For example, a burning or sore tongue and pale gums are common symptoms of iron, folic acid, and vitamin-B12-deficiency anemias, while sickle-cell anemia may present itself as an unnatural paleness of the mucous membranes."

(秦荻辉. 精选科技英语阅读教程[M]. 西安:西安电子科技大学出版社,2008.)

正如本文开头所述那样,本文告诉读者"口腔检查可以发现的疾病多达40多项,其中部分疾病甚至性命攸关。"

本文的翻译难度不大。翻译时,需查明注意医学术语的意思,理清信息之间的关系,用地道的汉语流畅地写出原文的意思即可。

1. 词汇层

lower jaw 下颚;blood pressure 血压;physician 内科医师;heart attack 心脏病发作;symptom 症状;cavity 洞(文中译作"蛀牙");whitish patch 白色斑点;the National Institute of Dental Research 美国国立牙科研究院;the American Dental Association 美国牙医协会;barometer 晴雨表;the Illinois Masonic Medical Center 伊利诺伊州共济会医疗中心;gastrointestinal tract【解剖】胃肠道,胃肠管;the American Cancer Society 美国癌症协会;oral cancer 口腔癌;pathologist【病理】病理学家;malignancy 恶性(肿瘤等);periodontist 牙周病学家;canker sore 口腔溃疡,口疮;herpes【皮肤】疱疹;trench mouth 溃疡性牙龈炎;nemia【医】贫血,贫血症;folic acid【生化】叶酸,维生素B;sickle-cell anemia【医】镰状细胞性贫血;mucous membrane 黏膜。

2. 句子层

E. g. Reluctantly, he took time from work to visit the nearest dentist.

医学语篇英译汉时,译者同样须心怀英汉差异,用汉语的表达法表意。我们中国人说"寻医问药""就医""谨遵医嘱"等,英译汉时,汉语译文应尽量采用这样的说法。如本句中的"visit the nearest dentist"中的"visit"不宜译为"访问""拜访",而要根据上下文,可将"visit the nearest dentist"译为"前往最近的牙医处就诊"。本句可译为:"尽管罗杰斯很不情愿,他仍从工作中抽出时间前往最近的牙医处就诊。"

E. g. If you deal with the younger age group, then you see a lot of the acute infections such as herpes and trench mouth, which is due to bacteria and causes open sores between the teeth.

在科技英语文章中,"形合"的英语要求句中、句间信息关系要明示,所以一个英语长句写作往往一气呵成,逻辑关系明示,有利于英语读者获取信息;汉语倾向于有短句表意,对于英语长句汉译,第一印象是短句,即将 1 个英语长句断为 2~3 个汉语短句。本句中的定语从句"...,which is..."构成字眼超过 8 个,一般断为汉语单句。本句可译为:"在年龄比较小的患者人群中,可发现由细菌感染引起的疱疹、溃疡性牙龈炎等多种急性感染疾病。这些疾病可能会导致牙齿疼痛。"

E. g. For example, a burning or sore tongue and pale gums are common symptoms of iron, folic acid, and vitamin-B12-deficiency anemias, while sickle-cell anemia may present itself as an unnatural paleness of the mucous membranes.

本句中医学术语较多,翻译时须谨慎,将术语翻译准确。"present itself as"中的"present"在此处译为名词的"表现",读来会顺畅很多。本句可译为:"例如,舌头灼热感、舌头酸痛和牙龈苍白,这些是因缺乏铁、叶酸和维生素 B12 而造成的常见贫血症,而镰状细胞性贫血症的表现可能是口腔黏膜呈现看起来不自然的苍白色。"

【参考译文】

牙医可从您口腔获取的信息

【口腔检查可以发现的疾病多达 40 多项,其中部分疾病甚至性命攸关。】

大卫·罗杰斯的下颚疼痛越来越严重,开始影响他的注意力。尽管罗杰斯很不情愿,他仍从工作中抽出时间前往最近的牙医处就诊。

牙医彻底检查了罗杰斯的口腔,甚至测量了他的血压。取下听诊器后,牙医对罗杰斯说:"我能听出您的心跳很不规律。现在您或许并没有什么异常的感觉,但我认为在采取下一步治疗措施之前,首先应该检查一下您的心脏。"

罗杰斯表示反对,因为除下颚疼痛,他并没有其他不适感。在牙医的坚持下,罗杰斯最终做出让步,同意前往同一所医院大楼中的心脏科接受医生检查。罗杰斯埋怨称:"只要让我的下颚不再疼痛,我听您的安排。"

结果表明,这次检查救了罗杰斯的命。医生检查发现,罗杰斯已处于心脏病发作的早期阶段。幸运的是,在心脏病加重之前,罗杰斯可通过接受治疗阻止心脏病发作。罗杰斯是幸运的。牙医清楚地知道下颚疼痛可能是心脏病发作的症状之一。几个月后,罗杰斯心怀感激再次拜访那位牙医,这次只是为了表示感谢。罗杰斯告诉牙医:"我这条性命极可能是您给救的。"

尽管露西·帕克很少患有蛀牙,但她仍坚持定期接受牙科检查。

在一次牙科检查中,牙医在露西的口腔中发现白色斑点,结果表明这个斑点是癌症症状。根据她的病情,她毫无疑问患了癌症。这次牙科检查真地救了露西的命。

大卫和露西的故事都真实存在。这两个故事说明定期接受牙医检查至关重要。六个月实施一次的牙医检查不仅可以提供有关牙科疾病的警告信息,而且还可提供有关身体疾病的重要信息。

美国国立牙科研究院(National Institute of Dental Research)患者护理部门负责人迈克尔·罗伯茨(Michael Roberts)医生声称:"癌症和心脏病是人类的两大杀手。牙医检查不仅仅可发现这两种疾病。根据美国牙医协会报告,口腔检查可以发现40多种严重疾病,其中包括糖尿病、贪食症、脑肿瘤和艾滋病等。"

伊利诺伊州共济会医疗中心(Illinois Masonic Medical Center)牙科部主任劳伦斯·科恩(Lawrence Cohen)声称:"口腔是身体健康最明显、最易得的晴雨表。口腔属于容易检查的身体区域,所以口腔就像是体内疾病的一面镜子。与胃肠道检查需要使用插管不同,口腔检查仅需要明亮的光线和牙科检查镜即可。"

美国癌症协会(American Cancer Society)估计,仅今年就有29 000多名患者被确诊患有口腔癌,其中9 500名患者会死去。佛罗里达州口腔病理学专家黛安·斯特恩(Diane Stern)声称:"如果您经常进行口腔检查,您若不幸患有口腔癌,在癌症早期阶段可能就会发现。确诊时恶性肿瘤越小,治愈的可能性就越大。"佛罗里达州牙周病学专家杰里·罗森鲍(Jerry Rosenbaum)医生同意这一观点,声称:"患者生存几率和癌症发现时间存在直接相关性。"

牙医还可以检查出不会直接威胁生命的多种健康问题。科恩医生指出:"口腔溃疡是人们最常见的疾病之一。在年龄比较小的患者人群中,可发现由细菌感染引起的疱疹、溃疡性牙龈炎等多种急性感染疾病。这些疾病可能会导致牙齿疼痛。牙医还可能发现其他诸如口腔疼痛、味品不佳等疾病。"

科恩医生还指出:"在早期阶段,口腔检查还可检查出贫血等血液疾病。例如,舌头灼热感、舌头酸痛和牙龈苍白,这些是因缺乏铁、叶酸和维生素B12而造成的常见贫血症,而镰状细胞性贫血症的表现可能是口腔黏膜呈现看起来不自然的苍白色。"

15.2.2 汉译英

上呼吸道感染

上呼吸道感染简称上感,又称普通感冒,是包括鼻腔、咽或喉部急性炎症的总称。广义的上感不是一个疾病诊断,而是一组疾病,包括普通感冒、病毒性咽炎、喉炎、疱疹性咽峡炎、咽结膜热、细菌性咽-扁桃体炎。狭义的上感又称普通感冒,是最常见的急性呼吸道感染性疾病,但发生率较高。成人每年发生2~4次,儿童发生率更高,每年6~8次。全年皆可发病,冬春季较多。

别称:普通感冒

英文名称:upper respiratory tract infection

就诊科室:呼吸内科

多发群体:老、幼、免疫功能低下或患有慢性呼吸道疾病

病因

上呼吸道感染有 70%～80% 由病毒引起。包括鼻病毒、冠状病毒、腺病毒、呼吸道合胞病毒、埃可病毒、柯萨奇病毒等。另有 20%～30% 的上感由细菌引起。细菌感染可直接感染或继发于病毒感染之后，以溶血性链球菌为最常见，其次为流感嗜血杆菌、肺炎球菌、葡萄球菌等，偶或为革兰阴性细菌。

各种导致全身或呼吸道局部防御功能降低的原因，如受凉、淋雨、气候突变、过度疲劳等可使原已存在于上呼吸道的或从外界侵入的病毒或细菌迅速繁殖，从而诱发本病。老幼体弱，免疫功能低下或患有慢性呼吸道疾病的患者易感。

临床表现

根据病因和病变范围的不同，临床表现可有不同的类型：

1. 普通感冒

俗称"伤风"，又称急性鼻炎或上呼吸道卡他，多由鼻病毒引起，其次为冠状病毒、呼吸道合胞病毒、埃可病毒、柯萨奇病毒等引起。

起病较急，潜伏期 1～3 天不等，随病毒而异，肠病毒较短，腺病毒、呼吸道合胞病毒等较长。主要表现为鼻部症状，如喷嚏、鼻塞、流清水样鼻涕，也可表现为咳嗽、咽干、咽痒或灼热感，甚至鼻后滴漏感。发病同时或数小时后可有喷嚏、鼻塞、流清水样鼻涕等症状。2～3 天后鼻涕变稠，常伴咽痛、流泪、味觉减退、呼吸不畅、声嘶等。一般无发热及全身症状，或仅有低热、不适、轻度畏寒、头痛。体检可见鼻腔黏膜充血、水肿、有分泌物，咽部轻度充血。

并发咽鼓管炎时可有听力减退等症状。脓性痰或严重的下呼吸道症状提示合并鼻病毒以外的病毒感染或继发细菌性感染。如无并发症，5～7 天可痊愈。……

(https://baike.baidu.com/item/%E4%B8%8A%E5%91%BC%E5%90%B8%E9%81%93%E6%84%9F%E6%9F%93/8256336? fr＝aladdin&fromid＝502565&fromtitle＝%E6%84%9F%E5%86%92)

这是网上介绍上呼吸道感染的一篇医学专业文章的前半部分。这里节选了上呼吸道感染的"病因"和部分"临床表现"。本篇包含很多医学术语，涉及医学原理，翻译难度大。

1. 词汇层

鼻腔 nasal cavity；咽 pharynx；喉部 larynx；病毒性咽炎 viral pharyngitis；喉炎 laryngitis；疱疹性咽峡炎 herpangina；咽结膜热 pharyngeal conjunctival fever；细菌性咽-扁桃体炎 bacterial pharyngitis and amygdalitis；鼻病毒 rhinovirus；冠状病毒 coronavirus；腺病毒 adenovirus；呼吸道合胞病毒 respiratory syncytial virus；埃可病毒 echovirus；柯萨奇病毒 coxsackievirus；溶血性链球菌 hemolytic streptococcus；流感嗜血杆菌 haemophilus influenzae；肺炎球菌 pneumococcus；葡萄球菌 staphylococcus；革兰阴性细菌 gramnegative bacteria；上呼吸道卡他 upper respiratory tract catarrh；咽痒 pharyngeal itching；黏膜充血 mucosal hyperemia；咽痛 pharyngalgia；流泪 lacrimation；味觉减退 hypogeusia；声嘶 hoarseness；鼻腔黏膜充血 nasal mucosa hyperemia；分泌物 secreta；咽鼓管炎 salpingitis；鼻病毒 rhinoviruses；并发症 complication。

2. 句子层

E. g. 广义的上感不是一个疾病诊断，而是一组疾病，包括普通感冒、病毒性咽炎、喉炎、疱疹性咽峡炎、咽结膜热、细菌性咽-扁桃体炎。

本句中"不是……而是……"对应英语中的"not... but..."句式,其中"是"应转换为英语的"refer to",而不是"is";"包括……"后罗列了一串并列信息,翻译时一一罗列即可。本句可译为:"In a broad sense, URTI does not refer to a single disease, but to a group of diseases, including the common cold, viral pharyngitis, laryngitis, herpangina, pharyngeal conjunctival fever, bacterial pharyngitis and amygdalitis."

E.g. 狭义的上感又称普通感冒,是最常见的急性呼吸道感染性疾病,多呈自限性,但发生率较高。

本句中"又称……""是……""呈……""发生率……"的逻辑主语都是"上感",这是汉语松散成性的表达方式。仔细阅读分析后,我们得知本句讲了2层意思:①对"上感"定性;②对"上感"评价,宜译为两句。

当一句话中有两个以上主谓结构时,英语中往往以其中一个主谓结构支撑起该句的架子,承载着该句中相对来说最重要的信息,让其他主谓结构设法附着在该主谓结构上,这样英语信息主次分明。本句可译为:"**An upper respiratory infection**, also known, in the narrow sense, as the common cold, **is** the most frequently **occurring** of the infectious acute respiratory diseases. Although it is self-limiting, its incidence rate is higher than that of other upper respiratory infections."

译文中第一句划线的黑体字是本句的核心信息,其他信息是次要信息,附着在主句之上。

E.g. 细菌感染可直接感染或继发于病毒感染之后,以溶血性链球菌为最常见,其次为流感嗜血杆菌、肺炎球菌、葡萄球菌等,偶或为革兰阴性细菌。

原文含有3个逗号和1个句号。为准确表意,翻译的关键在于明示逗号前后的逻辑关系,尽可能译为英语长句。原文医学术语较多,这不可怕,一一罗列即可,尽量顺译。笔者将"细菌感染可直接感染或继发于病毒感染之后"译为英语主句,将其他信息附着在主句这个主谓结构上,注意"以……最常见"和"其次为……"的英语处理方法。本句可译为:"Bacterial infections can be either direct or secondary to the viral infection, with hemolytic streptococcus being the most common type, followed by haemophilus influenzae, pneumococcus and staphylococcus, or, occasionally, gram-negative bacteria."

E.g. 各种导致全身或呼吸道局部防御功能降低的原因,如受凉、淋雨、气候突变、过度疲劳等可使原已存在于上呼吸道的或从外界侵入的病毒或细菌迅速繁殖,从而诱发本病。

在"各种导致全身或呼吸道局部防御功能降低的原因"中,"原因"前的定语中可解读出1个动宾结构:"导致功能降低",在此可以为定语从句,用来修饰"原因"。"从而诱发本病"表示结果,译作状语。本句可译为:"Various phenomena which weaken the defense function either of the whole body or of the local respiratory tract, such as catching a cold, exposure to rain, abrupt change of climate and overfatigue, may lead to the rapid proliferation of viruses or bacteria already existing in the upper respiratory tract or those entering it from the outside, thus triggering the disease."

E.g. 起病较急,潜伏期1~3天不等,随病毒而异,肠病毒较短,腺病毒、呼吸道合胞病毒等较长。

这是汉语典型的表述方式:信息呈竹节状,信息单位短,随呈随释,层层展开,不断向前。英语信息具黏连性,信息主次分明,关系明晰。翻译成英语时,正确地搭建句子结构对翻译质量起决定性作用。充分挖掘本句中信息之间的关系后,本句可译为:"With its comparatively rapid onset, the disease has an incubation period of 1 to 3 days, shorter than an enterovirus but longer than an adenovirus or a respiratory syncytial virus, depending on the type of virus."从译文可见,笔者将"潜伏期1~3天不等"译为英文译文的主句,让其他信息附着在这个主谓结构之上,层次分明,表意清晰。

E. g. 主要表现为鼻部症状,如喷嚏、鼻塞、流清水样鼻涕,也可表现为咳嗽、咽干、咽痒或灼热感,甚至鼻后滴漏感。

本句虽长,但不难翻译。为使译文行文更地道,须将汉语动词的"表现"译为英语名词"manifestation"。并列的医学术语多,须翻译准确。本句可译为:"Its main manifestation is nasal symptoms, such as sneezing, nasal obstruction and runny nose, as well as cough, dry throat, pharyngeal itching or burning, and even postnasal drip."

E. g. 一般无发热及全身症状,或仅有低热、不适、轻度畏寒、头痛。体检可见鼻腔黏膜充血、水肿、有分泌物,咽部轻度充血。

在明晰了信息之间的逻辑关系后,我们得知,可将"或"前后解读为让步关系,用"although"连接。"体检可见"后列举的病状,翻译时一一列举前须生成"symptoms"这个概括性词语,统领以下信息,这符合英语从总到分的表达习惯。本句可译为:"Although generally there is no fever or constitutional symptoms, there can be a low-grade fever, discomfort, and headache. A physical examination will reveal such symptoms as nasal mucosa hyperemia, edema, secreta and mild pharyngeal hyperemia."

E. g. 并发咽鼓管炎时可有听力减退等症状。脓性痰或严重的下呼吸道症状提示合并鼻病毒以外的病毒感染或继发细菌性感染。

本句翻译若有困难,在于"有听力减退等症状"在英文中的句式搭建方式和"提示""合并鼻病毒以外"等词汇层的转换上。"并发咽鼓管炎时"中的"……时",须解读为"in case of",因为不总是"并发咽鼓管炎"。本句可译为:"In cases of concurrent salpingitis, symptoms of hyperacusis will appear. Purulent sputum or severe lower respiratory tract symptoms will suggest a virus or secondary bacterial infection other than the combined rhinoviruses."

E. g. 体检可见喉部水肿、充血,局部淋巴结轻度肿大和触痛,可闻及喉部的喘鸣音。

原文中"体检可见"表述既武断,又肯定。实际上,科学地讲,人和人之间体质上有差异,患上感后,存在本句中列举的这些症状仅仅是可能的,不是必然的,所以不能如汉语原文那样表述得那样肯定。本句可译为:"A physical examination can determine whether there are symptoms such as laryngeal edema and hyperemia, a mild enlargement of the local lymph node, and haphalgesia, as well as whether laryngeal wheezing rales can be heard."

【参考译文】

Upper Respiratory Tract Infection

An upper respiratory tract infection (URTI), also known as the common cold, is the general term for an acute inflammation of the nasal cavity, pharynx or larynx. In a broad sense, URTI does not refer to a single disease, but to a group of diseases, including the common cold, viral pharyngitis, laryngitis, herpangina, pharyngeal conjunctival fever, bacterial pharyngitis and amygdalitis. An upper respiratory infection, also known, in the narrow sense, as the common cold, is the most commonly infectious acute respiratory disease with a high rate of incidence. Although it is self-limiting, its incidence rate is higher than that of other upper respiratory infections. For adults, the incidence rate is 2 – 4 times annually and for children, 6 – 8 times annually. Although attacks of the common cold can occur all year round, they are more frequent in winter and spring.

Alternative name: Common cold

English name: Upper respiratory tract infection

Clinic department: Department of Respiratory Medicine

Susceptible groups: The elderly, children, and those with low immune function or who suffer from chronic respiratory disease

Pathogenesis

Generally, 70% – 80% of upper respiratory tract infections are caused by viruses which include the rhinovirus, coronavirus, adenovirus, respiratory syncytial virus, echovirus, coxsackievirus, etc. In addition, another 20% – 30% of URTIs are caused by bacteria. Bacterial infections can be either direct or secondary to the viral infection, with hemolytic streptococcus being the most common type, followed by haemophilus influenzae, pneumococcus and staphylococcus, or, occasionally, gram-negative bacteria.

Various phenomena which weaken the defense function either of the whole body or of the local respiratory tract, such as catching a cold, exposure to rain, abrupt change of climate and overfatigue, may lead to the rapid proliferation of viruses or bacteria already existing in the upper respiratory tract or those entering it from the outside, thus triggering the disease. The elderly, the weak, children and patients with low immune function or suffering from chronic respiratory disease are susceptible.

Clinical manifestation

According to the pathogenesis and extent of the disease, the clinical manifestation can vary in type:

1. Common cold

The common cold is also known as acute rhinitis orupper respiratory tract catarrh. It is most often caused by a rhinovirus, and with lesser frequency by a coronavirus, a respiratory syncytial virus, a echovirus, or a coxsackievirus, etc..

With its comparatively rapid onset, the disease has an incubation period of 1 to 3 days, shorter than for an enterovirus but longer than for an adenovirus or a respiratory syncytial virus. Its main manifestation is nasal symptoms, such as sneezing, nasal obstruction and runny nose, as well as cough, dry throat, pharyngeal itching or burning, and even postnasal drip. At the time of the attack or within several hours, symptoms such as sneezing, nasal obstruction, watery running nose etc. appear. symptoms at the same with the onset Two to three days later, thick nasal mucus appears, accompanied by pharyngalgia, lacrimation, hypogeusia, difficulty in breathing, hoarseness, etc. Although generally there is no fever or constitutional symptoms, there can be a low-grade fever, discomfort, and headache. A physical examination will reveal such symptoms as nasal mucosa hyperemia, edema, secreta and mild pharyngeal hyperemia.

In cases of concurrentsalpingitis, symptoms of hyperacusis will appear. Purulent sputum or severe lower respiratory tract symptoms will suggest a virus or secondary bacterial infection other than the combined rhinoviruses. If no complications exist, the patient can recover within 5-7 days...

练 习 题

一、请将下列文本译为汉语。

Telemedicine Comes Home

Medicine: Telemedicine permits remote consultations by video link and even remote surgery, but its future may lie closer to home.

Few places on earth are as isolated as Tristan da Cunha. This small huddle of volcanic islands, with a population of just 269, sits in the middle of the South Atlantic, 1,750 miles from South Africa and 2,088 miles from South America, making it the most remote settlement in the world, so it is a bad place to fall ill with an unusual disease, or suffer a serious injury. Because the islands do not have an airstrip, there is no way to evacuate a patient for emergency medical treatment, says Carel Van der Merwe, the settlement's only doctor. The only physical contact with the outside world is a six to seven-day ocean voyage, he says. "So whatever needs to be done, needs to be done here."

Nevertheless, the islanders have access to some of the most advanced medical facilities in the world, thanks to Project Tristan, an elaborate experiment in telemedicine. This field, which combines telecommunications and medicine, is changing as technology improves. To start with, it sought to help doctors and medical staff exchange information, for example by sending X-rays in electronic form to a specialist. That sort of thing is becoming increasingly common. "What we are starting to see now is a patient-doctor model," says Richard Bakalar, chief medical officer at IBM, a computer giant that is one of the companies in Project Tristan.

A satellite-internet connection to a 24-hour emergency medical center in America enables Dr. Van der Merwe to send digitized X-rays, electrocardiograms (ECGs) and lung-function tests to experts. He can consult specialists over a video link when he needs to. The system even enables cardiologists to test and reprogram pacemakers or implanted defibrillators from the other side of the globe. In short, when a patient in Tristan da Cunha enters Dr. Van der Merwe's surgery, he may as well be stepping into the University of Pittsburgh medical center. It is a great comfort to local residents, says Dr. Van der Merwe, knowing that specialist consultations are available.

Most of the technology this requires is readily available, and it was surprisingly simple to set up, says Paul Grundy, a health-care expert at IBM. The biggest difficulty, he says, was to install the satellite-internet link. In theory, this sort of long-distance telemedicine could go much further. In 2001 a surgeon in New York performed a gall-bladder removal on a patient in Paris using a robotic-surgery system called Da Vinci. Although that was technologically impressive, it may not be where the field is heading.

Home is where the technology is.

For advances in telemedicine are less to do with the tele than with the medicine. In the long term, it may be less about providing long-distance care to people who are unwell, and more about monitoring people using wearable or implanted sensors in an effort to spot diseases at an early stage. The emphasis will shift from acute to chronic conditions, and from treatment to prevention. Today's stress on making medical treatment available to people in remote settings is just one way telemedicine can be used and it is merely the tip of a very large iceberg that is floating closer and closer to home...

(韩孟奇.科技英语阅读[M].上海:上海交通大学出版社,2012.)

二、请将下列文本译为英语。

天然产物是自然界的生物在进行生命活动过程中合成的代谢产物。源于天然产物的药物已经被广泛用于治疗多种重大疾病,如心血管疾病、恶性肿瘤、免疫疾病和传染性疾病等。相较于化学合成的小分子药物,天然产物在结构新颖性、生物相容性和功能多样性等方面具有明显的优势,并且在长期进化过程中得到自然筛选优化。

在新药研发和临床用药中天然产物及其衍生物占有很大比例。据统计,在1939年至2016年间,美国食品及药物管理局(FDA)批准的上市药物中,有相当数量含有天然产物的分子片段(50%以上),甚至直接来源于天然产物。因此,利用天然产物进行药物设计是现代创新药物产生的有效途径。活性天然产物与细胞内靶标相互作用是其发挥作用的基础。

基于天然产物新药发现的第一个关键步骤就是靶点的确定。在药物研发的起始阶段明确药物分子的靶点,有助于深入研究其作用机制,并尽早发现其可能存在的毒副作用,进而从结构上进行有针对性的改造,降低药物研发成本。

在生物体内,包括天然产物在内的小分子化合物(Compound)通过结合大分子靶标(Target),在基因组(Genome)、转录组(Transcriptome)、蛋白质组(Proteome)和代谢组(Metabolome)等各个层面引起表型(Phenotype)变化。由于蛋白质是细胞功能的主要执行者，大分子靶标在多数情况下是与天然产物相互作用的蛋白质，即靶点蛋白。根据小分子-靶点-表型三者之间的逻辑关系，天然产物靶点鉴定的策略可以分为两大类：首先，以小分子天然产物为起点筛选靶标蛋白，或被称作逆向策略（reverse strategy），包括化学蛋白质组学（Chemical proteomics）、化学基因组学（Chemical genetics）和生物物理学（Biophysics）等方法；其次，根据天然产物引起的表型变化或与之相关的已知信号通路和作用网络推测并确定天然产物的靶标，也称作正向策略（forward strategy），主要包括差异基因组学/蛋白质组学分析、细胞形态分析等方法。本文将通过近20年来的代表性实例介绍这些靶标鉴定的方法策略。

（周怡青，肖友利.活性天然产物靶标蛋白的鉴定[J].化学学报，2018，76(3):177－189.）

第 16 章 航空语篇翻译

16.1 航空语篇的分类

语域(Register)是语言使用的场合或领域的总称(360 百科)。不同语域的语篇在体裁、题材、词汇、句法、文体等特征方面存在差异。

根据语域,航空语篇大致可以分为以下类型:
(1)民航服务语篇。
(2)飞行语篇。
(3)航空工程及维修语篇。
(4)航空领域各分支学科内的专业语篇、学术论文等。

根据文体形式,航空语篇大致可以分为以下类型:
(1)实用文本:比如各飞机制商制定的各型号飞机的飞行手册、机组人员的维修手册、设备清单、维修大纲等。
(2)公文文本:比如国际民航组织、美国联邦航空局、欧洲航空安全局等机构发布的适航通告和行业规范。
(3)法规文本:比如国际民航组织的公约、附件、议事规则、会议决议、欧洲联合航空管理规定及美国《联邦法规汇编》相关文件及条款等。
(4)科技论文文本:比如航空领域中的研究人员所发表的国内外期刊、会议论文等。

16.2 航空语篇的特点

除了具有一般科技语篇所具有的的文体特点之外,航空语篇在词汇、句式、文体等语言层面有特殊之处。

(1)词汇特点:航空英语词汇有大量的缩略词、合成词、派生词、拼合词、剪切词、外来词、新造词,翻译实践中可以依据构词法来推测、记忆和翻译相关词汇。

E. g. Madar＝Malfunction Analysis Detection And Recording 故障分析与记录;MAIN＝Material Automated Information Network 材料自动信息网路;burning-rate 燃烧速率(田建国,2018);autopilot 自动驾驶仪;avionics＝aviation＋electronics 航空电子学。

(2)句法特点:名词化结构多,被动语态多,非限定性结构和介词短语多,从句和长句多。

E. g. All communications from the flight compartment are handled through the audio selector panels which provide source selection, and volume control for microphone and audio output.

机组舱所有的对外联络工作由音频选择器面板处理,音频选择器面板可以对麦克风和音频输出进行声道选择和音量控制。

E. g. The constant 32.2, is derived from the fact that a freely falling object is accelerated by the force of gravity 32.2 feet per second each second it falls.

常数 32.2 来源于以下这个事实,即自由落体物体在下落的过程中,由于受到重力的吸引,每秒钟以 32.2 英尺/秒的速度加速。

(3)文体特点:语言正式,语气客观公正,逻辑严密,句法规范,信息准确。

16.3 航空语篇翻译实例

16.3.1 英译汉

Power-plant

(1) An aircraft engine, or power-plant, produces thrust to propel an aircraft. Reciprocating engines and turboprop engines work in combination with a propeller to produce thrust. All of these Power plants also drive the various systems that support the operation of an aircraft.

(2) Reciprocating Engines.

Most small aircraft are designed with reciprocating engines. The name is derived from the back-and-forth, or reciprocating movement of the pistons which produces the mechanical energy necessary to accomplish work.

(3) Reciprocating engines can be classified as:

1) Cylinder arrangement with respect to the crankshaft: radial, In-line, V-type, or opposed.

2) Operating cycle: two or four.

3) Method of cooling: liquid or air.

(4) Radial engines were widely used during World War II and many are still in service today. With these engines, a row or rows of cylinders are arranged in a circular pattern around the crankcase. The main advantage of a radial engine is the favorable power-to-weight ratio.

(5) In-line engines have a comparatively small frontal area but their power-to-weight ratios are relatively low. In addition, the rearmost cylinders of an air-cooled, in-line engine receive very little cooling air, so these engines are normally limited to four or six cylinders. V-type engines provide more horsepower than in-line engines and still retain a small frontal area. The continued improvements in engine design led to the development of the horizontally-opposed engine which remains the most popular reciprocating engines used on

smaller aircraft. These engines always have an even number of cylinders, since a cylinder on one side of the crankcase "opposes" a cylinder on the other side. The majority of these engines are air cooled and usually are mounted in a horizontal position when installed on fixed-wing airplanes. Opposed-type engines have high power-to-weight ratios because they have a comparatively small, lightweight crankcase. In addition, the compact cylinder arrangement reduces the engines frontal area and allows a streamlined installation that minimize aerodynamic drag.

(6) The main parts of a spark ignition reciprocating engine include the cylinders, crankcase, and accessory housing. The intake/exhaust valves, spark plugs, and pistons are located in the cylinders. The crankshaft and connecting rods are located in the crankcase. The magnetos are normally located on the engine accessory housing.

(7) In a four-stroke engine the conversion of chemical energy into mechanical energy occurs over a four stroke operating cycle. The intake, compression, power, and exhaust processes occur in four separate strokes of the piston.

1) The intake stroke begins as the piston starts its downward travel. When this happens, the intake valve opens and the fuel/air mixture is drawn into the cylinder.

2) The compression stroke begins when the intake valve closes and the piston starts moving back to the top of the cylinder. This phase of the cycle is used to obtain a much greater power output from the fuel/air mixture once it is ignited.

3) The power stroke begins when the fuel/air mixture is ignited. This causes a tremendous pressure increase in the cylinder, and forces the piston downward away from the cylinder head, creating the power that turns the crankshaft.

4) The exhaust stroke is used to purge the cylinder of burned gases. It begins when the exhaust valve opens and the piston starts to move toward the cylinder head once again.

(8) Even when the engine is operated at a fairly low speed the four-stroke cycle takes place several hundred times each minute. In a four-cylinder engine, each cylinder operates on a different stroke. The continuous rotation of a crankshaft is maintained by the precise timing of the power strokes in each cylinder. The continuous operation of the engine depends on the simultaneous function of auxiliary systems, including the induction, ignition, fuel, oil, cooling, and exhaust systems.

(沈星.航空科技英语[M].北京:北京理工大学出版社,2015.)

本文涉及往复式发动机的类型和工作原理,文字难度适中,篇章结构单一,内容易懂。这是一篇信息类的短文,翻译时首先须确保信息准确,同时须做到译文通顺、得体,其次要求篇章结构与原文保持一致。

1. 词汇层

本文有一些与飞机发动机相关的专业术语,翻译时力求准确,可通过网络上专业词典如《蓝天航空词典》(该词典可从网络语料库中抓取含有某词条的多个例句,这样译者在具体的上下文中可解读出该词条的含义)或纸质的航空类大词典来解决,切不可望文生义或给出模棱两可的译文,否则会给委托人造成一定损失或误导读者。

reciprocating engine and turboprop engine 往复式发动机和涡轮螺旋桨发动机；the crankshaft 曲轴；operating cycle 运行周期；the crankcase 曲轴箱；power-to-weight ratio 推重比；the horizontally-opposed engine 水平对置型发动机；accessory housing 发动机附件壳；intake/exhaust valves 进气/排气阀；spark plugs 火花塞；magneto 磁电机；four-stroke engine 四冲程发动机；four-cylinder engine 四缸发动机。

2. 句子层

E. g. Most small aircraft are designed with reciprocating engines.

本句中短语"are designed with"可译成"设计装载",增加"装载"一词,用来补充原文短语的语义,也可以看作"装载"一词是对"with"的翻译,"with"一词起衔接功能,而汉语是意合语言,缺少衔接手段,往往需用词汇来补充(下文还会有类似的处理手法)。本句可译为:"大多数小型飞机设计装载往复式发动机。"

E. g. The name is derived from the back-and-forth, or reciprocating, movement of the pistons which produces the mechanical energy necessary to accomplish work.

本句中定语从句"which produces...",同前一句一起构了本段的信息推进模式——连续型主位推进模式,所以翻译时要注意句内各成分之间顺序的排列,既要忠于原文,又要符合汉语表达习惯。翻译时,"the name"可译成"往复发动机的名称",与前一句译文同词重复,"the back-forth, or reciprocating movement of pistons"的译文"活塞的前后往复运动"也重复使用两次,一是因为汉语缺少前后文照应的指代手段,二是因为通过同词重复可以把前后逻辑表达清楚。本句可译为:"往复式发动机的名称来源于活塞的前后往复运动,活塞的前后往复运动产生了推动飞机的机械能。"

E. g. With these engines, a row or rows of cylinders are arranged in a circular pattern around the crankcase.

本句句首介词短语"with these engines"在原文中起强调作用,与后文逻辑关系紧密,而汉语一般以话题作为句子的开头,因此笔者把它处理成单句的主语,句尾的方式状语"in a circular pattern around the crankcase"中"circular"和"around"都有"围绕"的意思,译文中保留一个即可,这是作为意合语言的汉语措词相对节省的优点表现。本句可译为:"这类发动机是由一排或者多排气缸围绕曲轴箱布置而成。"

E. g. The continued improvements in engine design led to the development of the horizontally-opposed engine which remains the most popular reciprocating engines used on smaller aircraft.

本句中有三个名词短语,结构平行,信息量大,这是航空英语或科技英语典型的特点。笔者根据汉语的表达习惯,把前两个含有"improvement"和"development"具有动作含义的名词短语译成动词短语,并且把整个原文主句拆分成两个汉语小分句。第三个名词短语"the most popular reciprocating engines used on smaller aircraft"里面有一个后置定语,翻译时要提到中心词的前面。另外,原文的定语从句是对主句的逻辑总结,译文也就顺势作结,不做句序调整。本句可译为:"随着发动机设计的不断创新,人们开发出了水平对置型发动机。这种发动机至今仍是小型飞机上最流行的往复式发动机型号。"

E. g. In addition, the compact cylinder arrangement reduces the engine's frontal area and allows a streamlined installation that minimize aerodynamic drag.

本句中有四个名词短语、三个动词和一个定语从句,翻译重点在于通过这三个动词和定语从句来确定四个名词短语的前后顺序和逻辑关系,然后按照汉语的行文规范重新排列。这句话的逻辑可以这样表达:第一个名词短语导致第二个名词短语表达的结果,第二个名词短语又导致第三个名词短语表达的结果,而第三个名词短语又是直接导致第四个名词短语的原因,层层递进,一环扣一环。本句可译为:"另外,紧凑的气缸排列减小了发动机的最大截面,可以进行流线型安装,这样气动阻力降到最低。"

E. g. In a four-stroke engine the conversion of chemical energy into mechanical energy occurs over a four stroke operating cycle. The intake, compression, power, and exhaust processes occur in four separate strokes of the piston.

本句中有多个名词,来说明四冲程发动机的工作原理。这两句话用了同一个词"occur"做谓语,起搭建句子架构的作用。本句整体意思和成分之间的逻辑关系若要表达清楚,就需要做拆分、调整原短语顺序、转换词性、转化词义、增加词汇等工作。

本句可译为:"在一个四冲程发动机里,燃油的化学能转化为机械能,这个转化工作通过运行一个四冲程循环实现。进气、压缩、做功(燃烧)和排气过程是活塞运动的四个不同的冲程。"第一句话先把原文主语短语拆开,"conversion"名词转成动词,增加"燃油"一词,补充信息,谓语动词"occurs"译为"实现"更加具体,"operating"转化成动词并且顺序提前,使汉语译文通顺,且符合发动机原理。为了符合汉语句式规则,原文信息浓缩的简单句,主语和谓语部分分别以一个汉语短句来表述,为了衔接两个小分句,第二个分句开头重复使用"转化"这个词。

【参考译文】

动力装置

(1)飞机的发动机或动力装置产生推力推动飞机。往复式发动机和涡轮螺旋桨发动机同螺旋桨配合起来产生推力。所有这些动力装置同时也驱动各种支持飞机运行的辅助系统。

(2)往复式发动机。

大多数小型飞机设计装载往复式发动机。往复式发动机的名称来源于活塞的前后往复运动,活塞的前后往复运动产生了推动飞机的机械能。

(3)往复式发动机可分为以下三类。

1)根据气缸排列和曲轴的位置关系分为辐射式、直排式、V型或者对置式发动机。

2)根据运行周期数分为两冲程或四冲程发动机。

3)根据制冷方法分为液冷或者气冷发动机。

(4)辐射式发动机在第二次世界大战期间广泛应用,而且很多在今天还在使用。这类发动机是由一排或者多排气缸围绕曲轴箱布置而成。辐射式发动机的主要优势是其良好的推重比。

(5)直排式发动机有相对较小的最大截面,但是它的推重比相对较低。另外,气冷式直排发动机最后面的气缸只能接收到很少的制冷气流,因此这类发动机只能安装四个或六个气缸。V型发动机比直排式发动机提供更多马力,且仍然保留了较小的最大截面。随着发动机设计的不断创新,人们开发出了水平对置型发动机,这种发动机至今仍是小型飞机上最流行的往复式发动机型号。这类发动机的气缸总是有偶数个,因为曲轴箱一侧的气缸和另一侧的气缸对置。对置型发动机绝大多数是气冷式的,当安装于固定翼飞机时,通常安装在水平位置。对置式发动机的曲轴箱相对小而轻,所以它的推重比高。另外,紧凑的气缸排列减小了发动机的最大截面,可以进行流线型安装,这样气动阻力降到最低。

(6)火花塞点火往复式发动机的主要部件包括气缸、曲轴箱和附件壳。进气/排气阀、火花塞和活塞位于气缸内部。曲轴和曲轴连杆位于曲轴箱内部。磁电机通常位于发动机附件壳内部。

(7)四冲程发动机里,燃油的化学能转化为机械能,这个转化工作通过运行一个四冲程循环实现。进气、压缩、做功(燃烧)和排气过程是活塞运动的四个不同冲程。

1)进气冲程从活塞向下移动开始。此时,进气阀门打开,燃油空气混合物被吸入气缸。

2)进气阀门关闭后,活塞朝气缸顶部移动,压缩冲程开始。一旦油气混合体点燃,这个阶段的循环会产生更大的动力输出。

3)点燃油气混合气体后,做功(燃烧)冲程开始,此时气缸内压力大大增加,活塞离开气缸头向下运动,产生旋转曲轴的动力。

4)排气冲程用于清除气缸中燃烧过的气体。排气阀门打开后,活塞再次朝气缸顶部移动,排气冲程开始。

(8)即使发动机运行速度相对较低,四冲程循环也会每分钟发生几百次。在四缸发动机中,每个气缸同时运行不同的冲程。曲轴的连续旋转是由每个气缸的做功(燃烧)冲程的精确计时来维持的。发动机的连续运行依赖进气、点火、燃油、润滑、制冷和排气等辅助系统的同步运行。

16.3.2 汉译英

发电机试验台操作规程

禁止非指定操作人员操作发电机试验台

一、操作前检查

　　1. 电源电压:(380±38)V,三相电压应平衡。

　　2. 应安全可靠的安装被测件、连接工装和连接线。操作台所有控制开关应打到"OFF"。

　　3. 接地良好。

二、操作

　　1. 接通主电源。

　　2. 起动拖动齿轮箱的润滑冷却循环油系统,并观察其油压值,油压达到规定范围0.1~0.3 MPa,可通过调节手柄实现。

　　3. 接通操作台电源,观察转向开关挡位是否正确,再将被测件拖动到要求的转速范围发电、加速。

　　4. 试验结束后应先卸去负载,然后将转速慢慢降至零。

5. 关断操作台电源,关断电源配电柜的电源开关。

三、注意事项

1. 试验中不要长时间接触被测件,以免被烫伤。

2. 测试过程中如有异常情况发生,一定要先将发电、励磁开关打到"OFF",再将转速慢慢降至零。

四、维护、保养

1. 试验台保持表面洁净。

2. 保持转接盘干净,工作时要给连接轴涂少量润滑脂。

3. 拖动台在使用过程中应注意其振动情况,当出现异常时,应停车检查。

4. 点检内容：

(1)齿轮箱润滑油干净、油量正常；

(2)空载试运转中,油路畅通、回油量正常、各接头处没有漏油现象；

(3)空载试运转中,齿轮箱及电机运转正常,无异常声响、无异常振动；

(4)各种仪器、仪表显示正常,操作系统灵敏、可靠。

5. 精度检查项目：

(1)齿轮箱输出主轴跳动量≤0.02 mm。

(2)齿轮箱输出主轴相对安装盘止口跳动量≤0.05 mm。

6. 保持润滑油、滤油器清洁,必要时添加或更换润滑油。

7. 切断电源,对电气箱、操作台、负载箱、动力柜进行除尘处理,拧紧螺钉。

8. 检查各种开关是否齐全、灵敏,指示灯是否明亮,必要时进行调整和更换,检查校验仪表是否符合指标要求,必要时进行修理和调整。

本文属于航空维修领域里的实用性文体,操作规程由许多条款组成,对操作人员进行指导,通篇是命令语气,语言简洁、信息准确、易于操作。译文应该忠实于原文的文体特点,做到每一条信息无误、言简意赅、易于理解和操作。

1. 词汇层

通过专业的网络词典或纸质词典,或请教专业人员解决专业术语的翻译问题。

发电机试验台 Electric Generator Test Bed；非指定操作人员 Unauthorized operator；电源电压 Power supply voltage；三相电压 Three-phase voltage；被测件 Device Under Test；连接工装和连接线 connection fixture and cables；操作台 the operating console；齿轮箱的润滑冷却循环油系统 the lubricating and cooling circulation oil system which drives gearbox；负载 load；电源配电柜 power distribution board；励磁开关 field switch；转接盘 adapting disk；连接轴 the connecting shaft；拖动台 driving platform；齿轮箱 gear box；空载试运转 no-load trail run；滤油器 oil filter；电气箱 electrical box；负载箱 load box。

2. 句子层

E.g. 禁止非指定操作人员操作发电机试验台。

本句是整个规程的前提,是一个无主句,这是汉语在禁令等祈使句中常见的句式。缺省的主语是所有的操作人员,所以在规程中无需写出。然而,一个完整的英语句子一般不能缺少主语(除非下文中各个条款中主语不言自明,用短语的效果更好)。"非指定操作人员"其实质语义是"未经授权人员",可译为"unauthorized operator",这样更符合本文的语域。译文中增加了主语,很自然地采用被动结构,这也是英文禁令中常用的形式。本句可译为:"Unauthorized operator are prohibited from operating the generator test bed."

E. g. 电源电压:(380±38)V,三相电压应平衡。应安全可靠的安装被测件、连接工装和连接线。操作台所有控制开关应打到"OFF"。接地良好。

这三条操作前检查的事项,谓语部分在英译文用"should be"(而不是"must be")来翻译原文"应"这个字表达。本句可译为:"Power supply voltage:(380±38)V, Three-phase voltage should be balanced. The installment of DUT (Device Under Test), connection fixture and cables should be safe and reliable. All control switches on the operating console should be turned to 'OFF'. All the equipment should be properly grounded."另外,"安装被测件、连接工装和连接线"本是动词短语,笔者在译文中将其转化成名词短语,充当句子主语,这样符合英文的句法规范。

E. g. 起动拖动齿轮箱的润滑冷却循环油系统,并观察其油压值,油压达到规定范围 0.1～0.3 MPa,可通过调节手柄实现。

本句中"拖动齿轮箱的润滑冷却循环油系统"呈现偏正结构,"的"前后是修饰和被修饰关系。其中前置定语"拖动齿轮箱"是动宾结构,译为英语时常译为定语从句。所以"拖动齿轮箱的润滑冷却循环油系统"可译为"the lubricating and cooling circulation oil system which drives gearbox"。"并观察其油压值,油压达到规定范围 0.1～0.3 MPa,可通过调节手柄实现"。这三个短句的逻辑关系在汉语中隐藏掉了,这是意合汉语常见的做法,翻译时译者须解读并明示逻辑关系。因此,可增加"make sure"和"by"以标明句子的逻辑关系,同时省略"实现"的译文。本句可译为:"Start the lubricating and cooling circulation oil system which drives gearbox, and observe the oil pressure value. Make sure the oil pressure reaches the prescribed range 0.1～0.3 MPa, by adjusting the handle."

E. g. 齿轮箱润滑油干净、油量正常;空载试运转中,油路畅通、回油量正常、各接头处没有漏油现象;空载试运转中,齿轮箱及电机运转正常,无异常声响、无异常振动;各种仪器、仪表显示正常,操作系统灵敏、可靠。

以上四句是点检的四项内容,每一条中都有"正常"一词,汉语的"正常"是个模糊词汇,翻译时应根据上下文做不同的解读和翻译,"油量正常"和"回油量正常"中"正常"一词在英文中用"normal"表达,取"正常、标准"之意,而"电机运转正常"和"仪器、仪表显示正常"中的"正常"分别可译为"as they should"和"accurately",取其有"准确无误"和"毫无偏差"之意。同时,为了加强效果,每一条增加了一个表示祈使语气的短语。本句可译为:"See that the lubricating oil in gear box is clean and oil mass normal; In no-load trial run, check that the oil circuit is smooth, the amount of scavenge oil is normal, and that there is no leakage in any of the joints; In no-load trail run, make sure that the gear box and the motor are operating as they should, with no abnormal sounds and vibrations; See that the various instruments, meters accurately display and that the operating system is sensitive and reliable."

E. g. 检查各种开关是否齐全、灵敏。

"齐全"一词,如果直译为"complete",即信息表达得模不清晰,所以笔者意译为"be equipped",这样更符合语境;同时,增加一个表示祈使语气的短语,以示强调。本句可译为:"Check to see whether the various switches are fully equipped and sensitive"。

【参考译文】

Operation Procedures forthe Electric Generator Test Bed

UnauthorizedOperator Are Prohibited From Operating the Generator Test Bed

Ⅰ. Pre-operation check

1. Power supply voltage: (380±38)V, Three-phase voltage should be balanced.

2. Installment of the DUT (Device under Test): the connection fixture and cables should be safe and reliable. All control switches on the operating console should be turned to "OFF".

3. All equipment should be properly grounded

Ⅱ. Operating steps

1. Turn on the main power supply.

2. Start the lubricating and cooling circulation oil system which drives the gearbox, and observe the oil pressure value. Make sure the oil pressure reaches the prescribed range 0.1~0.3 MPa, by adjusting the handle.

3. Switch on the operating console's power, observe whether the gear range of the change-over switch is correct, and then drag the DUT to within the range of the prescribed rotation speed to generate electricity and then accelerate.

4. After the test, remove the load, and then reduce the rotation speed to zero.

5. Turn off the operating console's power and switch off the power switch on the power distribution board.

Ⅲ. Caution

1. Avoid prolonged contact with the DUT during the experiment so as not to be burned.

2. If there are any anomalies in the experiment, be sure to turn the electricity-generating switch and the field switch to "OFF", and then gradually reduce the rotation speed to zero.

Ⅳ. Maintenance

1. Keep the test bed surface clean.

2. Keep the switch disk clean, and apply a small amount of grease to the connecting shaft.

3. In the course of operation, pay attention to the vibration of the driving platform and when anomalies appear, stop the machine and check.

4. Spot inspection instructions:

(1) See that the lubricating oil in the gear box is clean and the oil mass normal;

(2) In a no-load trial run, check that the oil circuit is smooth, the amount of scavenge

oil is normal, and that there is no leakage in any of the joints;

(3) In a no-load trail run, make sure that the gear box and the motor are operating as they should, with no abnormal sounds or vibrations;

(4) See that the various instruments and meters accurately display and that the operating system is sensitive and reliable.

5. Accuracy inspection items:

(1) The amount of Gearbox output shaft run-out should be ≤0.02mm.

(2) The amount of Gearbox output shaft run-out relative to the spigot of the mounting plate stop should be ≤0.05mm.

6. Keep the lubricating oil and the oil filter clean; replace or add oil if necessary.

7. Turn off the power and clean the dust from the electrical box, the operating console, the load box and the power supply cabinet and then tighten the screws.

8. Check to see whether the various switches are fully equipped and sensitive, and that the light is bright. Adjust and replace them if necessary; check whether the calibration instruments conform to the requirements of the index parameter. Repair and adjust them if necessary.

练 习 题

一、请将下列文本译为汉语。

Oil System

Reciprocating engines use either a wet-sump or a dry-sump oil system. In a wet-sump system, the oil is located in a sump, which is an integral part of the engine. In a dry-sump system, the oil is contained in a separate tank, and circulated through the engine by pumps.

The main component of a wet-sump system is the oil pump, which draws oil from the sump and routes it to the engine. After the oil passes through the engine, it returns to the sump. In some engines, additional lubrication is supplied by the rotating crankshaft, which splashes oil onto portions of the engine.

An oil pump also supplies oil pressure in a dry-sump system, but the source of the oil is located external to the engine in a separate oil tank. After oil is routed through the engine, it is pumped from the various locations in the engine back to the oil tank by scavenge pumps. Dry-sump systems allow for a greater volume of oil to be supplied to the engine, which makes them more suitable for very large reciprocating engines.

The oil pressure gauge provides a direct indication of the oil system operation. It ensures the pressure in pounds per square inch (psi) of the oil supplied to the engine. Green indicates the normal operating range, while red indicates the minimum and maximum pressures. There should be an indication of oil pressure during engine start.

Refer to the *AFM/POH* for manufacturer limitations.

The oil temperature gauge measures the temperature of oil. A green area shows the normal operating range and the red line indicates the maximum allowable temperature. Unlike oil pressure, the changes in oil temperature occur more lowly. This is particularly noticeable after starting a cold engine, when it may take several minutes or longer for the gauge to show any increase in oil temperature.

Check oil temperature periodically during flight especially when operating in high or low

ambient air temperature. High oil temperature indications may signal a plugged oil line, a low oil

quantity, a blocked oil cooler, or a defective temperature gauge. Low oil temperature indications may signal improper oil viscosity during cold weather operations.

(沈星.航空科技英语[M].北京:北京理工大学出版社,2015.)

二、请将下列文本译为英语。

油源车操作规程
禁止非指定人员操作油源车

一、操作前检查项

1.电源电压:(380±38)V,三相电压应平衡。

2.操作台所有控制开关应打到"OFF"。

二、操作

1.打开电源,先将加热管打到"ON",给润滑油加热升温。

2.当油温升至或接近要求值时,才能启动回油泵,并调节润滑油流量至规定范围。

3.启动拖动台,将被测件转速升至工作转速,调节流量和油压,使其符合规定的要求。如流量和油压有一项不符合要求,则不能继续进行试验。

4.被测件正常工作时,观察润滑油的压力、温度、流量达到规定值时,可断开加热开关打开冷水阀门,实行油温自动调节。

5.试验结束后,应先将被测件的转速慢慢降至零,再将润滑油流量调至零,然后等候1～2分钟,让被测件内的润滑油被充分抽干净后,关断回油泵电源。

6.关断油源车工作电源,切断总电源。

三、操作中注意事项

测试过程中如有异常情况发生,一定要先将被测件转速先降至零,再关断油源车。

四、维护、保养

1.操作台面板应经常擦拭,以保持表面洁净。

2.点检内容:

(1)每次实验之前检查油位,使供油油箱油位处于油标中位,加热油箱要用探油针探明离油箱高度2cm左右,达不到要求必须加油。

(2)每次实验之前检查过滤器是否干净,有脏物必须清除干净。

(3)油泵开启后工作有无异常振动现象,供油流量、压力及温度是否正常。

(4)各接头处是否有漏油现象。

(5)各种仪器仪表显示正常。

3.流量、压力、温度传感器按有关规定进行计量校验。

第17章　商务语篇翻译

"商务"概念有广义和狭义之分。"商务"广义上指的是"一切与买卖商品服务相关的商业事务",狭义上指的是"商业或贸易"(360 百科)。

随着我国改革开放向纵深发展,以"一带一路"建设为重点,全球范围的商务活动愈加频繁。商务活动涉及招商引资、技术引进、对外贸易、对外劳务承包和合同、国际金融、涉外保险、国际旅游、海外投资、国际运输(苑春明,姜丽,2013)等广阔领域。

频繁的商务活动离不开高品质的商务翻译。商务翻译要求既懂商务业务又懂外语、翻译能力强的专业人才。商务语篇翻译的翻译对象为与商务活动有关的文本。

17.1　商务语篇的分类

常见的商务语篇类型如下:
(1)商务合同。
(2)信件。
(3)通知。
(4)请示以及批复。
(5)电子邮件。
(6)会议纪要等。

17.2　商务语篇的特点

1. 格式固定

无论是合同、信件,还是通知等商务语篇,其写作有"套路"(固定的格式),易于学习,也易于使用。

2. 语气婉约

在现代社会,商务活动往往在公平原则和友好氛围下进行,伴随着礼貌、合理、大方、得体的言行,约定俗成的规矩和一定的商业礼仪。婉约的语气有利于促成商业活动。

3. 领域广阔

商务活动涉及金融、保险、旅游、贸易等多个领域,所以商务语篇也就涉及广阔的领域。

4. 用词专业

商务语篇(尤其是合同文本)有一些专业的用词。比如,合同文本中的"shall"表示"应当/必须"(不能用"should"表示),表示法律上必须履行的或应尽的义务,使用频率很高,语气很重;而"should"不表示法律义务,仅表示一般的义务或道义上的义务,意为"最好如此"。

17.3　商务语篇翻译实例

17.3.1　英译汉

Vat Fraud Crackdown Increases Liability of Website Marketplaces

Online marketplaces such as Amazon and eBay will be forced to police their own websites to prevent billions of pounds of VAT fraud, under a renewed government crackdown on the scam.

The move represents the second effort in as many years to close down an estimated £1.2bn-a-year tax fraud and follows a *Guardian* investigation as well as criticism of HM Revenue & Customs for being slow to act.

The scam involves foreign companies warehousing products in the UK and selling them, VAT free, via internet market places. All traders based outside the EU and selling goods online to UK customers should charge VAT if their goods are already in the UK at the point of sale.

The chancellor told parliament: "We are taking further action to address online VAT fraud, which costs the taxpayer £1.2bn per year, by making all online marketplaces jointly liable for VAT — (and) ensuring that sellers operating through them pay the right amount of VAT."

Online marketplaces will now be jointly liable with sellers for any unpaid VAT where the "online marketplace knew or should have known that the business should be registered for VAT in the UK".

Websites will also need to ensure that businesses operating on their site have displayed valid VAT numbers. Previous legislation, introduced last year, only imposed joint liability if the marketplace had not removed a seller 30 days after being informed by HMRC of a breach.

Richard Allen, founder of Retailers Against VAT Abuse Schemes, said: "This is a correction for the stupid bit of legislation they introduced last year which meant that Amazon and eBay did not have to act if they had not been notified of the fraud by HMRC first. If they do this properly it will kill this off. If I was Amazon or eBay I'd be petrified I'd be liable for the VAT if I hadn't checked the VAT numbers of the sellers adequately, which is the whole purpose of third-party liability."

Despite the huge amount of money involved in a seemingly unsophisticated crime, HMRC has been accused of being ineffectual in tackling it.

A report into the online VAT fraud, which was published by parliament's public accounts committee (PAC) last month, concluded: "HMRC has been slow to get to grips with the problem and is not yet doing enough to tackle it. Previous committee reports have highlighted the problem of online VAT fraud as a growing risk."

The report added: "HMRC has not named and shamed non-compliant traders and so far has not prosecuted a single seller for committing online VAT fraud."

Among the other measures designed to combat the fraud, the government also vowed to investigate the use of split payments, which involves a third party, such as a bank or marketplace, deducting VAT when they receive payment for goods. It will issue a report on split payments next month.

(https://www.theguardian.com/uk-news/2017/nov/24/vat-crackdown-increases-liability-of-website-marketplaces)

这是来自英国《卫报》的一篇文章,讲述了英国政府颁布打击增值税欺诈政策后,在线购物网站不得不监控自家网站的故事。

1. 词汇层

VAT(Value Added Tax)增值税;fraud 欺骗;HM Revenue & Customs(HMRC)英国税务海关总署;scam 骗局;parliament 议会;joint liability 连带责任;Retailers Against VAT Abuse Scheme 零售商反对增值税滥用计划;public accounts committee(PAC)公共账目委员会;prosecute 指控;split payment 分期付款。

2. 句子层

E. g. Online marketplaces such as Amazon and eBay will be forced to police their own websites to prevent billions of pounds of VAT fraud, under a renewed government crackdown on the scam.

汉语表意时往往状语先行,以作铺垫。"Online marketplaces such as..."中"Online marketplace"为概述,符合英语"先总后分(先总述,后分述)"的表达习惯,而汉语恰恰相反,要译为"诸如……等在线购物网站"。汉语表意时有一些概括性的表达法,如将"abortion"译为"堕胎行为"中的"行为",将"juvenile delinquency"译为"少年违法犯罪问题"中的"问题",均属于英译汉时增加的概括性的表达法,读来顺畅。本句可译为:"英国政府对增值税(VAT)欺诈采取新一轮打击行动,诸如亚马逊(Amazon)和易趣(eBay)等在线购物网站将不得不监控自家网站,以防止数十亿英镑的增值税欺诈犯罪状况继续恶化。"其中"犯罪状况继续恶化"是增加的信息,以保证译文读来顺畅。

E. g. The move represents the second effort in as many years to close down an estimated £1.2bn-a-year tax fraud and follows a Guardian investigation as well as criticism of HM Revenue & Customs for being slow to act.

翻译本句时,译者要理清信息之间关系。本句主语为"the move"(这一举措),谓语是"represent"(表明)和"follow"(字面义为"跟随",此处须结合上下文加以处理)。本句可译为:"英国税务海关总署(HMRC)因执法行动缓慢而饱受批评。《卫报》公布相关调查情况后,英国政府的这一举措表明,政府将再次着手处理已存在多年、每年估值高达 12 亿英镑的增值税欺诈问题。"

E. g. The chancellor told parliament："We are taking further action to address online VAT fraud, which costs the taxpayer £1.2bn per year, by making all online marketplaces jointly liable for VAT — (and) ensuring that sellers operating through them pay the right amount of VAT."

英语原文直接引语(括号中)信息为一个长句，一气呵成。汉译英时，要断为2~3句汉语句子，以迎合汉语用短句表意、隐藏逻辑关系词的语言性质。本句可译为："英国财政大臣向议会报告称：'我们正在采取进一步行动，以解决致使纳税人每年损失12亿英镑的增值税在线欺诈问题。我们要求在线购物网站承担卖家增值税纳税连带责任，确保通过其购物网站运营的卖家支付足额的增值税。'"可以看出，原文中"which costs the taxpayer £1.2bn per year"被译为前置定语，修饰"增值税在线欺诈问题"(online VAT fraud)，"问题"是增加的信息。原文"by making all online marketplaces..."被译为第二个汉语句子。

E. g. Online marketplaces will now be jointly liable with sellers for any unpaid VAT where the "online marketplace knew or should have known that the business should be registered for VAT in the UK".

英译汉时，译者一定要超越原文文本(不要被字面迷惑)，正确、透彻解读信息之间的关系。此处的"where the 'online marketplace...'"与其前信息实为因果关系，所以本句可翻译为："对于任何未按规定支付增值税的卖家，在线购物网站将承担连带责任，因为'在线购物网站了解或应当了解该卖家在英国应缴纳相应的增值税'。"

E. g. A report into the online VAT fraud, which was published by parliament's public accounts committee (PAC) last month, concluded："..."

翻译本句时，译者须将原文中定语从句"which was published by..."中的被动语态转换为汉语常用的主动语态表达习惯。本句可译为："英国议会公共账目委员会(PAC)在上个月发布的在线增值税欺诈报告中指出：'……'。"

E. g. Among the other measures designed to combat the fraud, the government also vowed to investigate the use of split payments, which involves a third party, such as a bank or marketplace, deducting VAT when they receive payment for goods.

本句中的"Among the other measures"字面义为"在其他措施中"，此处不可这样翻译。译者须明白这个短语与主句的关系，将此处最好译为"英国政府还采取其他措施……"。"which involves..."为定语从句，构成字眼超过了8个，一般须译为汉语单句，以迎合汉语用短句表意的做法。本句可译为："英国政府还采取其他措施打击增值税欺诈，承诺将调查分期付款规定的使用状况。按规定，若用户分期付款，银行或市场等第三方在收到货款时需扣除增值税。英国政府将于下月发布对分期付款的调查报告。"

【参考译文】

打击增值税欺诈政策增加在线购物网站责任

英国政府对增值税(VAT)欺诈采取新一轮打击行动，诸如亚马逊(Amazon)和易趣(eBay)等在线购物网站将不得不监控自家网站，以防止数十亿英镑的增值税欺诈犯罪状况继续恶化。

英国税务海关总署(HMRC)因执法行动缓慢而饱受批评。《卫报》公布相关调查情况后,英国政府的这一举措表明,政府将再次着手处理已存在多年、每年估值高达12亿英镑的增值税欺诈问题。

增值税欺诈涉及外国公司在英国的仓储商品及通过在线市场销售的免增值税商品。对于通过在线购物网站向英国国内客户销售商品且驻地处于欧盟区域之外的贸易商,若所售商品在售卖时已在英国国内,均应缴纳增值税。

英国财政大臣向议会报告称:"我们正在采取进一步行动,以解决致使纳税人每年损失12亿英镑的增值税在线欺诈问题。我们要求在线购物网站承担卖家增值税纳税连带责任,确保通过其购物网站运营的卖家支付足额的增值税。"

对于任何未按规定支付增值税的卖家,在线购物网站将承担连带责任,因为"在线购物网站知道或应当知道在英国注册企业应该缴纳增值税。"

此外,在线购物网站还需要确保卖家在其网站上标明有效的增值税号。去年英国颁布的法律规定,在线购物网站在收到英国税务海关总署有关在线卖家违法通知后,若未能在30天内移除该卖家,则需承担连带责任。

"零售商拒绝增值税滥用计划(Retailers Against VAT Abuse Schemes)"的创始人理查德·艾伦(Richard Allen)声称:"英国政府的这一举措是对去年愚蠢立法的修正。根据英国政府去年颁布的法律,如果事先未收到英国税务海关总署卖家违法的通知,亚马逊和易趣则无需对违法卖家采取行动。如果我是亚马逊或易趣的负责人,因未能详细检查零售商的增值税号而需要承受连带责任,我将压力倍增。这才是实施第三方连带责任的真正目的。"

尽管巨额增值税欺诈属于看似幼稚的欺诈犯罪,英国税务海关总署却因未能采取有效措施而备受指责。

英国议会公共账目委员会(PAC)在上个月发布的在线增值税欺诈报告中指出:"英国税务海关总署对在线增值税欺诈问题处理缓慢,现在做得依然不够。先前的委员会报告已强调在线增值税欺诈这一风险日益增加的问题。"

该报告还指出:"英国税务海关总署并未明确指出和惩罚违法的贸易商,迄今为止也未对任何卖家提出在线增值税欺诈指控。"

英国政府还采取其他措施打击增值税欺诈,承诺将调查分期付款执行情况。按规定,若用户分期付款,银行或在线购物网站等第三方在收到货款时需扣除增值税。英国政府将于下月发布对分期付款的调查报告。

17.3.2 汉译英

希望建立贸易关系函

××公司:

我们从商会那里看到贵公司的名称及地址,得知你们有兴趣建立进出口商品的业务联系。

如贵公司在本地尚无固定客户,希望考虑以本公司为交易伙伴。本公司原经营工业机械在本国的批发零售业务,由于最近在经营方面的变化,本公司在销售方面的政策也发生了变化。

笔者有多年的外贸经验,希望在世界各地建立适宜而持久的贸易关系。由于与生产厂家的长期直接联系,我们在许多行业中尤其是工业机械,是最有竞争力的。

我们也愿从贵国进口一两种优良产品,以有竞争力的价格在美国销售,以期能够持续、长期占领市场。

我们希望聆听贵公司的意见、要求及建议,以及如何才能使双方协力合作,互惠互利。此外,本公司愿意以收取佣金为条件充当贵公司在美国的采购代理。

恭候回音。

××公司

×年×月×日

(https://wenku.baidu.com/view/8c16f39cdd88d0d232d46a00.html)

这是一封商务信函,书写方希望与对方建立贸易关系。

1. 词汇层

商会 the chamber of commerce;工业机械 industrial machinery;批发零售 wholesale and retail;优良产品 high-quality product;占领市场 dominate market;协力合作 make a joint effort at cooperation;互惠互利 reciprocity and mutual benefit;佣金 commission;采购代理 purchasing agent。

2. 句子层

E. g. 希望建立贸易关系函

标题是文章的眼,直观地告诉读者文章内容。在信息爆炸时代的今天,这可以提高读者的阅读效率,所以标题很重要。本文标题"希望建立贸易关系函"表明这封信写作目的是"希望(与对方)建立贸易关系",所以这个标题可译为:"Letter of Intent:Establishing a Trade Relation"

E. g. 本公司原经营工业机械在本国的批发零售业务,由于最近在经营方面的变化,本公司在销售方面的政策也发生了变化。

"本公司经营……"中"经营",因主语为"公司",不宜译为"manage"或"run"(主语一般为人),而应理解为"从事"(engage in)。本句中第一个逗号前后为转折关系,应明示为"but"。本句可译为:"Our company engages primarily in the wholesale and retail of industrial machinery in our country, but due to a recent change in management, our company's policy regarding sales has also changed."

E. g. 由于与生产厂家的长期直接联系,我们在许多行业中尤其是工业机械,是最有竞争力的。

英文句子表达时有"语言单位向下使用"的趋势(即,若能用短语表达清晰,就不用句子;若能有单句表达清楚,就不用主从复合句表达)。比如,"我有一个红苹果"不应译为"I have an apple, which is red.",而应译为"I have a red apple."又如,"据悉,马哈蒂尔首先来到杭州,到访阿里巴巴总部,并与阿里巴巴董事局主席马云举行会面。杭州是马哈蒂尔此次中国行的第一站,……"。深挖信息之间的关系后,我们得知前一句中的"杭州"和第二句中的"杭州"为信息紧凑起见最好译在一起,译为"It is reported that he first visited the headquarters of Alibaba in Hangzhou, his first stop, and met with Ma Yun (Jack Ma), chairman of the board directors."从中可见,第二句"杭州是马哈蒂尔此次中国行的第一站"被译为"his first stop",句子变短语,语言单位变小。本句可译为:"With our long-term direct relations with manufacturers, we are the top competitor in many industries, especially industrial machinery."

E.g. 我们希望聆听贵公司的意见、要求及建议,以及如何才能使双方协力合作,互惠互利。

此处的"聆听"不应译为表动作的"listen to",而应译为表结果的"hear"。"以及……"与其前信息不应理解为并列关系,而是修饰与被修饰关系:是关于"如何才能使双方协力合作,互惠互利"方面(as to)的"意见、要求及建议"。因此,本句可译为:"We hope to learn your company's opinions and requirements and to hear suggestions as to how both parties can make a joint effort at cooperation in the interest of reciprocity and which will be to the mutual benefit of both."

【参考译文】

Letter of Intent: Establishing a Trade Relation

MM/DD/YY

×× company:

We received your company's name and address from the chamber of commerce and learned that you are interested in establishing business relations with respect to import and export commodities.

If you do not yet have any steady local clients, we hope that you might consider us as your trading partner. Our company engages primarily in the wholesale and retail of industrial machinery in our country, but due to a recent change in management, our company's policy regarding sales has also changed.

The writer has many years of experience in foreign trade and hopes to establish proper and permanent trade relations around the world. With our long-term direct relations with manufacturers, we are the top competitor in many industries, especially industrial machinery.

We are also willing to import one or two high-quality products from your country and sell them in the United States at competitive prices in order to continue to dominate the market for a long term.

We hope to learn your company's opinions and requirements and to hear suggestions as to how both parties can make a joint effort at cooperation in the interest of reciprocity and which will be to the mutual benefit of both. In addition, our company can also act as your company's purchasing agent in the United States on a commission basis.

Awaiting your reply.

××× company

练 习 题

一、请将下列文本译为汉语。

White Paper to Set out Industrial Strategy in Bid to Boost UK Productivity

Business secretary to announce creation of watchdog to monitor progress of white paper goals as biotech investment revealed.

Greg Clark believes the government has a role to play in putting UK companies at the vanguard of the new wave of technologies such as robotics and genomics. Photograph: Anthony Upton/PA

The government is to highlight five key areas where the UK needs to improve its performance when it reveals on Monday the details of a new industrial strategy designed to increase productivity.

Greg Clark, the business secretary, will announce the creation of an independent watchdog to monitor progress made in boosting innovation, upgrading infrastructure, increasing the level of workplace skills, ensuring that the strength of the City is reflected in funds for companies and spreading prosperity to all parts of the country.

A white paper to be published by Clark will also reveal that the government intends to set up long-term strategic deals in four sectors seen as having growth potential: construction, life sciences, automotive, and artificial intelligence.

New investment in the UK by the US-owned life sciences company Merck, known as MSD in Europe, creating 950 jobs, shows the benefits of the sort of partnership between Whitehall, the private sector and universities the government wants to create in its four sector deals, he will say.

"Our life sciences sector is one of the UK's fastest developing industries, with a turnover in excess of £64bn, employing 233,000 across the UK.

"MSD's commitment today, and the wider sector deal investment we have secured, proves the process outlined in the industrial strategy can give companies the confidence and direction they need to invest in the UK. It will ensure Britain continues to be at the forefront of innovation and represents a huge vote of confidence in our industrial strategy."

The white paper's focus on five core areas follows extensive consultation since the publication of a green paper at the start of the year. Clark believes the need for an industrial strategy has been made all the more pressing by last week's budget, in which the Office for Budget Responsibility halved its forecast of the UK's long-term productivity trend to 1%. The white paper will note that the need for the UK to seize available opportunities has been made more important by the Brexit vote.

(https://www.theguardian.com/business/2017/nov/27/clark-details-industrial-strategy-productivity-business-watchdog-white-paper)

二、请将下列文本译为英语。

商务合作协议书

甲方：
法定代表人：
住址：
邮编：
联系电话：
乙方：
法定代表人：
住址：
邮编：
联系电话：
甲乙双方本着平等自愿、互惠互利原则，就结成长期商务合作关系，经友好协商达成以下协议：
一、合作期限
本协议有效期为伍年。自×年×月×日起到×年×月×日止。
二、合作内容
甲方权利义务
1. 向乙方推荐合适的客户或项目。
2. 协助乙方促成与客户签约。
3. 如甲方自身还未与客户签约时，乙方的合作进程则听从甲方安排。
4. 甲方向乙方推荐的客户与乙方直接签约，甲方不负有任何责任。
乙方权利义务
1. 向甲方推荐合适的客户或项目。
2. 协助甲方促成与客户签约。
3. 为甲方推荐的客户提供最好的服务以及最优的价格。
4. 乙方同意以约定的结算时间和方式进行结算。

三、合作条件

1. 甲方向乙方或乙方向甲方推荐的客户成功与之签约,即视为推荐成功。

2. 成功推荐项目后,被推荐方向推荐方支付该项目实际营业额的2%做奖励。

3. 付款方式:自被推荐方首次收到客户服务费的第二个月开始支付,每月结算一次,每月5日前支付。推荐方则提供对应的正式发票。

四、违约责任

1. 合作双方在业务实施过程中,如因一方原因造成客户方商业信誉或客户关系受到损害的,另一方可立即单方面解除合作关系。同时,已经实现尚未结束的业务中应该支付的相关费用,受损方可不再支付,致损方则还应继续履行支付义务。

2. 双方在分配利润时,如任何一方对利润分配的基数、方式有异议的,可聘请会计师事务所进行审计。

五、补充变更

本协议在执行过程中,双方认为需要补充、变更的,可订立补充协议。补充协议具有同等法律效力。补充协议与本协议不一致的,以补充协议为准。

六、协议终止

1. 甲乙任何一方如提前终止协议,需提前一个月通知另一方。

2. 本协议期满时,双方应优先考虑与对方续约合作。

七、争议处理

如发生争议,双方应积极协商解决,协商不成的,受损方可向广州市人民法院提起诉讼。

八、协议生效

本协议经双方盖章后生效。本协议一式两份,甲乙双方各执一份,具有同等法律效力。

甲方:

代表签字:

日期: 年 月 日

盖章:

乙方:

代表签字:

日期: 年 月 日

盖章:

(https://wenku.baidu.com/view/09714024dd36a32d737581d1.html?from=search)

第 18 章 出国留学语篇翻译

改革开放 40 年来,我国取得了举世瞩目的成就。我国经济蓬勃发展,已跃居世界第二大经济体(仅次于美国),为我国取得更大进步打下了坚实的基础。"科技创新能力和水平快速提升,产出数量位居世界前列,产出质量大幅提高,已成为具有重要影响力的科技大国。"(来自搜狐)我国人民整体上实现了从贫穷到迈向小康生活的伟大转变,对未来美好生活充满了向往。由于我国经济的大幅增长,一定程度上出现"回国潮",但是还是有很多学子愿意出国留学深造,去长见识、长知识,为自己未来职业发展拓宽道路。

出国留学一般需要以下材料:①申请表;②推荐信;③个人陈述(Personal Statement);④个人简历(Resume);⑤在读证明和成绩单/学位证;⑥入学成绩;⑦申请信/附信(Cover Letter)等。不同学校之间要求略有差别,但大同小异。对我们中国人来说,这些材料往往是通过翻译将中文材料变为英文材料的。这些材料的翻译不难。

在下文中,笔者将分别讲解"推荐信""个人简历"和"个人陈述"的翻译。

18.1 书信的翻译

18.1.1 书信的构成要素

1. 称呼

要顶格、齐头写在第一行,后面加上冒号(中文)或逗号(英文),与下文隔开。

2. 问候语

要写在"称呼"的下一行,空两格(中文缩进 2 个汉字,英文缩进 4 个字母),或与"称呼"齐头书写,间隔一行,通篇相同。中文的问候语如"最近您好吗?",英文的问候语如"How are you doing recently?"。

3. 正文

就事说事,一般分为连接语、主体文、总括语三部分。

4. 祝颂语

这是一些表示致敬或祝福一类的话。中文如"此致""遥祝"等,一般要另起一行写,与之配套的"敬礼""健康""天天快乐"之类表祝福的话语,一般也要另起一行写。

5. 署名

署名前一般要加上"挚友""你的父亲""老公""教师"之类合适的称谓。

6. 日期

中文写信的日期一般写在"署名"的下一行靠右处，一般按年-月-日顺序写；英文要写在书信"称呼"的上一行靠右处，一般按月-日-年顺序写。

18.1.2 书信翻译实例

18.1.2.1 英译汉

May,2,2016

A Letter of Reference for Graduate Studies

Dear Prof. White,

　　I'm honored to recommend Jane Doe for admission to graduate studies at the Institute of Design. Since Jane will be visiting you soon, the following information should help introduce her as well as convey my whole hearted recommendation that she be admitted to your program.

　　My association with Jane — I've known Jane closely for three years, during which we've collaborated on many projects. Her role has been that of a consultant to our architectural firm. Jane and I have also conversed privately on theoretical and practical matters of style, aesthetics, and philosophies of design. I've always been impressed by her know-how — from how to design an arboretum to good principles of document design.

　　Jane's integrity — The hallmark of Jane's character is her honesty. This extends from those areas where we easily see it (in relationships and business transactions) to integrity of thought. Scientific, thorough, and meticulous, she approaches any analytical task with an exacting eye. This is the kind of care and concern I mean by integrity of thought.

　　Jane's scholarship and balanced education — As a student of the humanities, Jane pursued a course of study that could serve as a model for any general education curriculum. She knows not only Shakespeare but mathematics. She's as comfortable in a wood shop as at the opera.

　　Her professional manner — Jane Doe is one of the most pleasant persons you'll ever meet or work with. She's forthright, but neither intimidates nor intrudes. She's open, friendly, and authentic — the Jane you'll soon meet is the Jane you'll get, without any surprises.

　　Jane has my unqualified professional endorsement and my deep personal respect. Please feel free to call at any time if I can be of further assistance.

Your Sincerely,

All the best,

Jenny Pond

(https://wenku.baidu.com/view/39f7c031ee06eff9aef80732.html?from=search)

　　这是写信人詹妮·庞德向怀特教授推荐简·多伊入学攻读硕士学位的一封推荐信。信中介绍了詹妮与简的关系、简的诚实品格、简的学识与均衡教育、简的专业态度等方面情况。

翻译起来困难不大。

1. 词汇层

the Institute of Design 设计学院；collaborate（on）合作；aesthetics 美学；know-how 实际知识，专门技能；arboretum 植物园；integrity 诚实；hallmark 特点；business transaction 商业交易；meticulous 一丝不苟的；scholarship（此处）学识；curriculum 课程；forthright 直率的；intimidate 恐吓；endorsement 认可。

2. 句子层

E. g. I've known Jane closely for three years, during which we've collaborated on many projects.

本句较短，是由单句和定语从句构成的主从复合句。"know sb. closely"在汉语中没有现成的译法，"closely"表示"关系亲密"，不能字面和"know"（认识）拼合在一起。在保证翻译完整和译文通顺的情况下，可将"I've known Jane closely for three years,…"翻译为"我认识简并与之亲密相处已有三年，……"。本句后半部分为定语从句，修饰"three years"。虽然构成字眼没有超过8个（刚好8个），但若译作前置定语，汉语译文读来不通顺，所以翻译时断为汉语单句。本句可译为："我认识简并与之亲密相处已有三年，这期间我们合作过多个项目。"

E. g. She's as comfortable in a wood shop as at the opera.

在本句中，"wood shop"字面是"木工工厂"的意思。在本句所处的上下文中，不宜照字面翻译本句，否则会让人迷惑不解。本句可译为："无论是具体还是抽象工作，简都做得令人满意。"

E. g. She's forthright, but neither intimidates nor intrudes.

"intimidate"意思是"恐吓"，"intrude"意思是"侵扰"。这两词表示侵犯别人的程度不同。汉语列举这类信息时，要注意词序，表述时须按照汉语的表达法表意。本句可译为："简很直率，但不侵扰别人，更不威胁别人。"

【参考译文】

研究生学习推荐信

尊敬的怀特教授：

我非常荣幸地向您推荐简·多伊（Jane Doe）进入贵校设计学院攻读硕士学位。简不久之后将拜访您，我提供以下信息帮助您提前了解一下她，衷心希望她加入到您的研究项目中。

我与简的关系：我认识简并与之亲密相处已有三年，这期间我们合作过多个项目。简曾担任我们建筑公司的顾问，我们也曾私下探讨有关建筑风格、建筑美学、设计哲学等方面的理论和实践问题。从植物园设计到文件设计原则等各个方面，简渊博的知识给我留下了深刻印象。

简的诚实：在品行方面，简的最大特点是诚实。从容易观察到的领域（待人接物和商业交易）到思维完整性方面，处处体现她的诚实。简总是以科学、彻底、细致且严谨的态度完成所有分析任务。我所指的"思维完整性"正是这种对工作的关心和关注态度。

简的学识与均衡教育：作为人文学科的学生，简已按照通识教育课程模式完成了课程学习。她不仅懂莎士比亚而且还懂数学。无论是具体还是抽象工作,简都做得令人满意。

简的专业态度:简·多伊可能会是您遇到或一起工作过的最令人愉快的人之一。简很直率,但不侵扰别人,更不威胁别人。简性格开朗、为人友善且待人真诚——不久您见到她后就会了解她的这些性格特点,而不会感到任何惊讶。

简拥有我所不具备的专业水平,我个人对她怀有深深的敬意。如需我提供其他帮助,请随时与我电话联系。

此致

万事如意

詹妮·庞德

2016 年 5 月 2 日

18.1.2.2 汉译英

推 荐 信

四川大学高分子科学与工程学院高分子材料与工程专业 2006 级本科生"名字"是我"课程名"课班上的学生。该生在学习专业基础课的过程中积极主动,认真踏实。有严谨的学习作风和良好的学习习惯。在本课程取得了"成绩"分的好成绩,学习成绩优秀,在专业名列前茅。

该生不仅专业基础扎实,对科研有浓厚兴趣。虽然本科生的主要任务是学习专业基础知识,没有什么特别大的科研成果,但是该生参与申请本科生创新实验计划,在我的课题组参与创新实验"课题名称",在大三上学期进入实验室学习了"研究课题"的相关理论,并且熟悉了实验过程中的基本操作,在课余的时间来完成这个创新实验。

该生具有良好的英语能力,在各项英语考试中都获得了优异的成绩,为查阅英文以及以后撰写英文论文做好了铺垫。

该生全面发展,综合素质好,对科研有一定的兴趣,符合贵校的选拔条件。故特此推荐。

推荐人:×××

(https://wenku.baidu.com/view/f920dc6f27d3240c8447ef65.html)

这是网上一篇汉语推荐信,无称谓,无前后问候语,也无写信时间。

1.词汇层

高分子科学与工程学院 the College of Polymer Science and Engineering;积极主动,认真踏实 positive, conscientious, pragmatic and proactive;专业基础扎实 have a solid professional foundation;本科生创新实验计划 an undergraduate innovative experiment project;做好铺垫 lay a foundation for;全面发展 develop in an all-around way;综合素质好 have good general qualities;符合 in conformity with。

2.句子层

E. g. 四川大学高分子科学与工程学院高分子材料与工程专业 2006 级本科生"名字"是我"课程名"课班上的学生。

汉语表意时概念罗列往往从大到小,英语恰恰相反。本句中"四川大学高分子科学与工程学院高分子材料与工程专业 06 级本科生'名字'"即从大到小表意,译为英语时须转换为英语的从小到大表述习惯。当汉语前置的概念较短时，往往将前置语译为英语的同位语；当汉语前置的概念较长时，往往译为英语句子（单句或从句，视上下文来定）。本句"'名字'"前的前置语较长，宜译作句子。本句可译为："'Name', is an undergraduate (2006) majoring in polymer materials and engineering at the College of Polymer Science and Engineering at Sichuan University, and is a student in my 'course name' class."

E.g. 在本课程取得了"成绩"分的好成绩，学习成绩优秀，在专业名列前茅。

本句是对这位学生学习表现做出的评价，得高分、"成绩优秀"、"名列前茅"。英语表意时往往先总说，后分说，本句的翻译诀窍即在此。本句可译为："He/she has performed well in this course with a high score of "score points" and gets excellent grades and ranks among the top students in this major."

E.g. 虽然本科生的主要任务是学习专业基础知识，没有什么特别大的科研成果，但是该生参与申请本科生创新实验计划，在我的课题组参与创新实验"课题名称"，在大三上学期进入实验室学习了"研究课题"的相关理论，并且熟悉了实验过程中的基本操作，在课余的时间来完成这个创新实验。

从标点符号可见，本句仅包含一个汉语句子。翻译时第一印象是对本句翻译前要先断句。根据意群，应从"在大三上学期"前将句子断开，分开翻译。翻译时须明晰汉语原文逗号前后的逻辑关系。本句可译为："Although the main job of undergraduates is to acquire basic knowledge in their major field of study, thus they are not able to achieve especially significant results in scientific research, this student has participated in applying for an undergraduate innovative experiment project and is taking part in my "project name" research group. He/she joined the laboratory in the first semester of the junior year, and since has studied the theory relating to the research subject, become familiar with the basic operations of the experimental process, and completed the innovative experiment in his/her spare time."

E.g. 该生全面发展，综合素质好，对科研有一定的兴趣，符合贵校的选拔条件。

翻译本句的诀窍是要读透原文信息之间的逻辑关系。"符合贵校的选拔条件"的逻辑主语是其前的全部信息；即，"选拔条件"是"全面发展，综合素质好，对科研有一定的兴趣"。英语写作能力强的人知道，需将"符合贵校的选拔条件"处理为"which"引导的定语从句，其中引导词"which"往往指代主句整句信息内容。本句可译为："This student has developed in an all-around way, has good general qualities and has a pronounced interest in scientific research, all which is in conformity with the selection requirements (criteria) of your school."

【参考译文】

Letter of Recommendation

"Name", is an undergraduate (2006) majoring in polymer materials and engineering at the College of Polymer Science and Engineering at Sichuan University, and is a student in my "course name" class. In addition to being positive, conscientious, pragmatic and proactive in his/her studies, this student has a rigorous learning style and good study habits. He/she has performed well in this course with a high score of "score points" and gets excellent grades and ranks among the top students in this major.

This student not only has a solid professional foundation, but also a strong interest in scientific research. Although the main job of undergraduates is to acquire basic knowledge in their major field of study, thus they are not able to achieve especially significant results in scientific research, this student has participated in applying for an undergraduate innovative experiment project and is taking part in my "project name" research group. He/she joined the laboratory in the first semester of the junior year, and since has studied the theory relating to the research subject, become familiar with the basic operations of the experimental process, and completed the innovative experiment in his/her spare time.

This student has good English skills and has received excellent grades in all his/her English examinations, thus laying the foundation for referencing English and writing papers in English in the future.

This student has developed in an all-around way, has good general qualities and has a pronounced interest in scientific research, all which is in conformity with the selection requirements (criteria) of your school. Accordingly, I hereby recommend this student.

Recommended by ×××

18.2 个人简历的翻译

个人简历是求学、入职时需提供的必不可少的材料之一。

18.2.1 个人简历的构成要素及特点

18.2.1.1 个人简历的构成要素

1. 基本信息

往往包括姓名、性别、年龄、出生日期、民族、籍贯、住址、最终学历/学位、兴趣爱好、联系方式(邮箱地址、手机/座机号码)等。

2. 求职意向

在此写明意向职位、意向薪资、意向到岗时间等信息。

3. 学习经历

以时间为顺序罗列。尽量写2~3个学习阶段，一般要注明学习时间、专业和毕业院校等信息。比如，若博士毕业，可罗列本科、硕士和博士3个学习阶段。

4. 工作经历

工作经验很重要，这是非常重要的内容。在此，以时间为序，可详细罗列工作经历，一般包括工作时间、工种、工作地点等内容。

5. 能力、特长和荣誉

在此需如实描述自己的能力(尤其是与意向工作相关的能力)以及已取得的荣誉证书(包括外语、计算机或其他与工作相关的水平证书)，提交材料时须提交荣誉证书的复印件/原件，加以佐证。

6. 自我评价

务必客观,力求准确,篇幅以1~3句话为宜。

18.2.1.2　个人简历的特点

(1)时序性:罗列信息时以时间为序,方便读者阅读。

(2)客观性:提供的一切信息必须事实求是,不夸大,也不缩小。

(3)目的性:个人简历往往以求职或升学为目的,所以书写时,材料的取舍应以达到这个目的为标准,客观描述自己的"能力""特长"和已取得的"荣誉",尽量给读者留下"非你不录"的良好形象。

18.2.2　个人简历的翻译实例

18.2.2.1　英译汉

Name①

Street Address	Email Address
City, State, Zip Code	Phone Number
http://web. address/(*if applicable*)②	FAX (*if applicable*)

Objective③　　To obtain a _____ position in the area of _____ (*write this infinitive phrase to target the audience of this resume*)

Education④　　Name of Degree on Which You Are Working, Expected Month/Year of Graduation University

　　　　　　Overall GPA：X.XX/4.00　　Targeted GPA (*if beneficial*)：X.XX/4.00

　　　　　　Senior Design Project：Title of Senior Design Project

Relevant　Most Relevant Course　　　　　　Fourth Most Relevant Course

Courses　Second Most Relevant Course　　　Fifth Most Relevant Course

　　　　　Third Most Relevant Course　　　　Sixth Most Relevant Course

Experience⑤　Most Recent Position, Institution, Location (Month/Year – Month/Year)

　　　　　　Verb phrase that identifies key activity that you performed

---

① Creating a template or even a series of templates for a resume is perhaps impossible, because a resume, which is a summary of each individual, should be tailored for each individual. Given that, this template tries to give you some ideas about how to present yourself in a professional manner. The most important word grouping on a resume is the individual's name and should be the most prominent type on the page.

② Be sure to remove all comments in italics.

③ This infinitive phrase is often reworked by the author to target different audiences and purposes. In some situations, such as when a resume is attached to a proposal, this section is omitted.

④ For some people, presenting a GPA in the major or a GPA over the past two years puts that idividual in the best light.

⑤ On this template, most lists are kept to two, three, or four items. However, you might have good reason such as a large number of honors or awards to exceed that number. With lists, make sure that the items are listed in a parallel fashion. If the first item is a verb phrase, all items in that list should be verb phrases.

Second verb phrase that identifies key activity that you performed
Third verb phrase that identifies key activity that you performed (*if appropriate*)
Next Most Recent Position, Institution, Location(Month/Year – Month/Year)
Verb phrase that identifies key activity that you performed
Second verb phrase that identifies key activity that you performed
Third verb phrase that identifies key activity that you performed (*if appropriate*)
Third Most Recent Position, Institution, Location(Month/Year – Month/Year)
Verb phrase that identifies key activity that you performed
Second verb phrase that identifies key activity that you performed
Third verb phrase that identifies key activity that you performed (*if appropriate*)

Honors　Most Impressive Honor or Award
　　　　　Second Most Impressive Honor or Award
　　　　　Third Most Impressive Honor or Award
　　　　　Fourth Most Impressive Honor or Award

Activities　Most Impressive Activity
　　　　　Second Most Impressive Activity
　　　　　Third Most Impressive Activity
　　　　　Fourth Most Impressive Activity

Interests①　One outside interest, a second outside interest, perhaps a third outside interest

这是网上一篇模板式英文个人简历。提供的信息具体、全面,右侧提醒的信息很实用。翻译起来基本没有难度。翻译时,译者须细心,不要遗漏任何信息。

【参考译文】

姓名②

街道地址　　　　　　　　　　　　　　　　　　　　电子邮件
城市、国家、邮编　　　　　　　　　　　　　　　　电话号码
http://web. address/(如适用)③　　　　　　　　　　传真(如适用)
　目标　　　　为应聘＿＿＿＿＿＿领域内的＿＿＿＿＿＿职位(*使用不定式短语向本简历阅读人员提供希望获得的职位信息*)

---

①　The purpose of this section is to give an interviewer an ice-breaker at the beginning of an interview. For that reason, be prepared to talk about these activities when you interview. Also, if you are looking for a section to cut to fit your resume onto one page, then cut this section. Other possible sections to cut would be "Activities" and "Relevant Courses". In regard to space, if you do go to a second page. One means for doing so would be to write the resume in 12 point type, rather than 11 point type.

②　在简历中,应根据自己的实际情况提供扼要信息,因此创建(适合每个人的)简历模板或一系列模板也许是不可能完成的任务。鉴于此,本简历模板旨在帮助你以专业的方式介绍自己。简历中最重要的信息是个人姓名,应以最突出的形式体现出来。

③　一定要删除斜体文字表述的注释项。

教育经历①	你当前正在攻读的学位名称，以及大学毕业时间(月份/年份)	
	总平均成绩：x.xx/4.00　目标平均成绩(如有利)：x.xx/4.00	
	高级设计项目：高级设计项目名称	
相关课程	最相关的课程	第四相关的课程
	第二相关的课程	第五相关的课程
	第三相关的课程	第六相关的课程
工作经验②	最近从事的职位、机构、地理位置　　　　　(月/年—月/年)	
	描述你曾从事主要活动的动词短语	
	描述你曾从事主要活动的第二个动词短语	
	描述你曾从事主要活动的第三个动词短语(如适用)	
	最近从事的第二个职位、机构、地理位置　　　(月/年—月/年)	
	描述你曾从事主要活动的动词短语	
	描述你曾从事主要活动的第二个动词短语	
	描述你曾从事主要活动的第三个动词短语(如适用)	
	最近从事的第三个职位、机构、地理位置　　　(月/年—月/年)	
	描述你曾从事主要活动的动词短语	
	描述你曾从事主要活动的第二个动词短语	
	描述你曾从事主要活动的第三个动词短语(如适用)	
荣誉	印象最深的荣誉或奖励	
	第二个印象最深的荣誉或奖励	
	第三个印象最深的荣誉或奖励	
	第四个印象最深的荣誉或奖励	
活动	印象最深的活动	
	第二个印象最深的活动	
	第三个印象最深的活动	
	第四个印象最深的活动	
兴趣爱好③	第一个兴趣爱好、第二个兴趣爱好，可能存在的第三个兴趣爱好	

---

① 对于一些人员来说，提供专业平均分或过去两年课程的平均分，可以更好地展现自己。

② 在本模板中，该清单内容通常需列述两项、三项或四项，但如果你有更充分的理由(如获得多项荣誉或奖项)，则可以列述更多项。一定要并列列述各分项。如果第一项使用动词短语，则该清单随后各项都应使用动词短语。

③ 这部分旨在为面试官在面试开始时提供一个打开话题的机会。因此，你应做好谈论有关兴趣爱好话题的准备。另外，如果你希望在一张纸上写下简历的所有内容，可删除本部分。其他可删除的部分是"活动"和"相关课程"。如果你想用两张纸列述简历内容，那么你应尽量确保将第二页写满。要做到这一点，一个方法是用12(而不是11)号字体书写简历。

18.2.2.2 汉译英

个人简历文本

个人资料

  姓　名：　　　　　　　　　　政治面貌：

  性　别：　　　　　　　　　　学　历：

  年　龄：　　　　　　　　　　系　别：

  民　族：　　　　　　　　　　专　业：

  藉　贯：　　　　　　　　　　健康状况：

知识结构

  主修课：

  专业课程：

  选修课：

  实习：

专业技能

  接受过全方位的大学基础教育，受到良好的专业训练和能力的培养，在地震、电法等各个领域，有扎实的理论基础和实践经验，有较强的野外实践和研究分析能力。

外语水平

  1998年通过国家大学英语四级考试。1999年通过国家大学英语六级考试。有较强的阅读、写作能力。

计算机水平

  熟悉DOS、Windows 98操作系统和Office 97、Internet互联网的基本操作，掌握FORTRAN、Quick-Basic、C等语言。

主要社会工作

  小学：班劳动委员、班长。

  中学：班长、校学生会主席、校足球队队长。

  大学：班长、系学生会主席、校足球队队长，校园旗班班长。

兴趣与特长

  ☆喜爱文体活动、热爱自然科学。

  ☆小学至中学期间曾进行过专业单簧管训练、校乐团成员，参加过多次重大演出。

  ☆中学期间，曾是校生物课外活动小组和地理课外活动小组骨干，参加过多次野外实践和室内实践活动。

  ☆喜爱足球运动，曾担任中学校队、大学系队、校队队长，并率队参加多次比赛。曾获吉林市足球联赛(中学组)"最佳射手"称号并参加过1998嘉士伯北京市大学生足球联赛。

个人荣誉

  中学：×××优秀学生。×××优秀团员、三好学生、优秀干部。×××英语竞赛三等奖。

  大学：校优秀学生干部1996年度、1998年度三等奖学金与1997年度二等奖学金。

主要优点

★ 有较强的组织能力、活动策划能力和公关能力,如:在大学期间曾多次领导组织大型体育赛事、文艺演出,并取得良好效果。

★ 有较强的语言表达能力,如:小学至今,曾多次作为班、系、校等单位代表,在大型活动中发言。

★ 有较强的团队精神,如:在同学中,有良好的人际关系;在同学中有较高的威信;善于协同"作战。"

自我评价

活泼开朗,乐观向上,兴趣广泛,适应力强,勤奋好学,脚踏实地,认真负责,坚毅不拔,吃苦耐劳,勇于迎接新挑战。

求职意向

可胜任应用××××及相关领域的生产、科研工作。也可以从事贸易、营销、管理及活动策划、宣传等方面工作。

总的来说,书信、个人简历、个人陈述等实用型文本比信息型文本翻译起来要容易得多,主要基于以下原因:①实用型文本有比较固定的写作模式和模板,翻译时有些内容处理方法相同;②实用型文本内容更贴近真实生活,翻译的第一步,即阅读理解,根本就没有问题,所以翻译起来往往易于操作;③实用型文本(尤其是书信)也许是我们最熟悉的文体了,曾经无论是阅读还是写作都做了不少练习。笔译是目的语写作,翻译实用型文本可视作是用另一种语言重现原文信息的过程。

这篇汉语版个人简历翻译起来同样不难,但是译者须谨小慎微,注意一些细节之处的处理方法。

1. 词汇层

政治面貌 political status;民族 ethnicity/ethnic group;籍贯 place of birth/native place;选修课 optional/elective course;实习 internship;劳动委员 member of work team;校足球队队长 captain of the soccer team;单簧管训练 clarinet training;乐团 orchestra;文体活动 recreational and sports activity;野外实践 field practice;最佳射手 The Top Scorer;三好学生 merit student;公关能力 public relations ability;体育赛事 sporting event;文艺演出 artistic performance;团队精神 team spirit;人际关系 interpersonal relationship;威信 reputation;适应力强 highly adaptable;脚踏实地 down-to-earth;吃苦耐劳 endure hardship;营销 marketing;活动策划 activity planning;宣传 publicity。

2. 句子层

E. g. 接受过全方位的大学基础教育,受到良好的专业训练和能力的培养,在地震、电法等各个领域,有扎实的理论基础和实践经验,有较强的野外实践和研究分析能力。

这个句子有多个逗号,却包含多于1个意思,这是汉语有时不够科学的句读在汉语写作上的具体表现,翻译困难在于断句。从哪里将这句话断开呢?"在地震……"前表示受教育情况,后表示"理论基础""实践经验"和"能力"。如果将这些信息全部并列列出,会导致并列信息太长,读来枯燥,句式也没有节奏感。本句可译为:"Received a comprehensive basic university education and good professional skills training. I have a solid theoretical foundation and practical experience in the field of earthquake and electrical methods, a rich experience in field work and a strong ability of research analysis."

E. g. 活泼开朗,乐观向上,兴趣广泛,适应力强,勤奋好学,脚踏实地,认真负责,坚毅不拔,吃苦耐劳,勇于迎接新挑战。

成语是一种相习沿用的特殊固定词组,言简意丰,意义上具有整体性,结构上具有凝固性。成语使语言简洁,可增强修辞效果。由于从古汉语文本中沿习而来,汉语成语以四字居多。由于英汉差异,翻译时以翻译意思为主,无法做到字字对译。

本句中四字表达法多,一一罗列,翻译的困难在于如何理清信息层次而正确断句。断句的能力不但和译者对原文信息的理解能力有关,也跟译者对目的语的生成能力有关。本句中较多的四字表达法可译作相应的英语形容词,如"活泼开朗"可译作"outgoing","勤奋好学"可译作"studious",但是个别不行,比如"兴趣广泛""吃苦耐劳",所以翻译时要灵活处理,能译为英语形容词的可以并列起来,不能并列的信息设法附着在已生成的译文结构之上。本句可译为:"I am an outgoing, optimistic person with broad interests. I am highly adaptable, studious, down-to-earth, serious, responsible, firm and persistent; I am hardworking and can endure hardships; I am not afraid to meet new challenges."

【参考译文】

Personal Resume Text

Personal information

  Name:　　　　　　　　　　Political status:
  Sex:　　　　　　　　　　　Educational background:
  Age:　　　　　　　　　　　Department:
  Ethnicity:　　　　　　　　　Major:
  Place of birth:　　　　　　　Health status:

Knowledge structure

  Main course:
  Major course:
  Optional course/Elective course:
  Internship:

Professional Skills

Received a comprehensive basic university education and good professional skills training. I have a solid theoretical foundation and practical experience in the field of earthquake and electrical methods, a rich experience in field work and a strong ability of research analysis.

Foreign Language proficiency

In 1998, I passed the College English Test Band4. In 1999, I passed the College English Test Band 6. Have strong reading and writing abilities.

Computer skills

  Am familiar with DOS, Windows 98 operating systems and the basic operations of

Office 97 and the Internet, and have mastered FORTRAN, Quick-Basic and C language, etc.

Social work

Primary school: Committee member of work team, monitor.

Middle school: Monitor, chairman of university students' union, captain of the soccer team.

University: Monitor, chairman of student union of the department, captain of the soccer team, monitor of campus flag class.

Interests and specialties

☆ Like recreational and sports activities and love natural science.

☆ Received professional clarinet training during the period from primary school to middle school, served as a member of the school orchestra, and participated in significant performances many times.

☆ During middle school, served as the backbone member of the school biological extracurricular activities and geographical extracurricular activities group, and participated in the field practice and indoor practice activities for many times.

☆ Love soccer, served as the captain of the middle school team, department team and school team, and led team participation in competitions many times. Won the title of "The Top Scorer" of Jilin Football League (middle school group) and took part in the 1998 Carlsberg Beijing College Students' Football League.

Personal honors

Middle school: ×× × outstanding student, ×× × outstanding league member, merit student, outstanding cadre, ×× × third prize in English language competition.

University: School outstanding student cadre, third-class scholarship in 1996 and 1998, second-class scholarship in 1997.

Strong points

★ Strong organizing ability, activity planning ability and public relations ability. For example, I led and organized major sporting events and artistic performances many times while at the university, and achieved very good results.

★ Strong ability to express myself in language. For example, from primary school to now. I've made speeches many times "on important occasions" on behalf of the class, department and school.

★ Strong team spirit. For example, I have good interpersonal relationships with my classmates and enjoy a good reputation. I am good at cooperative "combat".

Self-evaluation

I am an outgoing, optimistic person with broad interests. I am highly adaptable, studious, down-to-earth, serious, responsible, firm and persistent; I am hardworking and can endure hardships; I am not afraid to meet new challenges.

Job intentions

I am competent in the application of... and production and scientific research in related fields. Engage in the trade, marketing, management, activity planning and publicity, etc.

(http://www.jianli-moban.com/n116c3.aspx)

18.3　个人陈述的翻译

也许是因为留学/求职申请人太多而无法为每位申请人提供面试机会，也许报考院校/意向单位觉得没有必要面试，个人陈述是招录官了解申请人的重要渠道和必要材料。

18.3.1　个人陈述的构成要素及特点

18.3.1.1　个人陈述的构成要素

一篇好的个人陈述文章应该包括以下内容（以留学为例）。

1. 选择本校/本专业/本课程的原因

有动机才有动力。申请者须清楚说明选择该校的原因，说明自己目前的学习基础，最好能提供未来的学习计划，让招生官相信"你"不但有现实"想法"，而且有"办法"（有能力从该校顺利毕业，迎接未来的挑战）。最好能展示出"你""独特"（异于他人）的一面，给招生官留下深刻印象。

2. 课程外的证明

在个人陈述中，最好能展示一个"立体"的自我，所以书写时还需展示学习之外自己的能力，如爱好、工作/学习经历、社会责任感（如参与慈善活动）、乐观的心态、积极的个性、组织领导力等。这些对塑造自我良好的形象帮助很大。

18.3.1.2　个人陈述的特点

一般来说，个人陈述有以下鲜明的特点。

1. 内容真实

在《论语·为政》中孔子说："言而无信，不知其可。"英语中有"Honesty is the best policy."（诚实乃上策）这样的谚语。在西方发达国家和我国都建有征信制度，将个人/企业的诚信纳入考评中，多实行一票否决制。因此，在个人陈述中，一定要真实陈述个人的相关情况，不得照抄模板内容，更不能生编硬造。

2. 文风朴实

招生官往往需要阅读大量的申请材料，疲惫不堪，简单、有力、重点突出的文章最能打动他们。因此，书写时，不要总是追求大词、表现煽情、强调"个性"，这些因素太多效果往往适得其反，而要用朴实的文风逻辑清晰地表达需要表达的内容。

3. 个性鲜明

书信、简历、个人陈述等实用型文本本身格式比较固定，给人"千篇一律"感，但是对申请者来说，申请者是在通过这些文本与招生官展开也许绝无仅有的一次"对话"，文本表现出的鲜明的个性有助于抓住招生官的眼球，往往是决定申请者能否胜出的关键。

18.3.2 个人陈述的翻译实例

18.3.2.1 英译汉

Personal Statement

Until college, I didn't know what my interest really was. In senior high school, what I learned was just for college entrance examination which was a big challenge for all Chinese students. I majored in economics by chance, while it must be one of the most intelligent choices that I had ever made. I fell in love with it gradually. I believe that the study of economics will be my lifelong career.

To be frank, I didn't know what economics was at the beginning. What I knew about it was many economists and their different theories. I even worried about whether things I learned were useful or not. One day, Miss Qi, my economic statistics teacher, asked us to solve an economic problem by what we had learned. It was the first time that I knew how to use the knowledge I had ever known. Only when you mastered a series of theories, could you find a meaningful problem. I had read a lot of literature before I decided to explore the problem of farmers' income which was related to farmers' daily life. Then I began to search the tool for processing data. It was not until you tried a lot that you could find a practical tool. At last, I found the factors that influenced the income of farmers. When I learned that my research results consistent with famous economist, I knew my effort was worthful. After that, I realized what I learned was the greatest science that could solve practical problems in human life. As long as you pay effort, could you move forward.

Now, thinking in economic way has become a part of my daily life. For example, the rise in oil prices for others maybe a good opportunity to visit filling stations. However, I do think how it will affect the international balance of payments and what monetary policy should the government adopt. Therefore, economics has become a part of my life that I can't live without it.

I am full of curiosity about everything that is very simple for others, but for me it's so interesting that I can think about it for hours. I think what is the smell after you dry the quilt, the principle of the zipper and so on. Every small thing that you can't imagine has chance to stimulate my interest. I will never give up until I find the answer. I have ever learned for 12 hours continuously to find how to use theeviews, which is a tool to analyze data that you collect. Economics is to me, what water is to fish.

I love the atmosphere in school where everything makes me feel comfortable that any other place can't. I want to become a teacher so that I can never leave school where I have stayed for 12 years. It's high time to make choices what to do next. Conditions permitting, I'm looking forward with great anticipation to the challenges that studying for a degree in economics will bring.

(https://wenku.baidu.com/view/dbeea737e45c3b3567ec8bfb.html? from＝search)

这是网上一篇文风朴实、情真意切的个人陈述文本，作者为中国人。翻译难度不大。

1. 词汇层

economic statistics 经济统计学；literature 文献；to process data 处理数据；filling station 加油站；balance of payment 收支平衡；monetary policy 货币政策；zipper 拉链。

2. 句子层

E. g. In senior high school, what I learned was just for college entrance examination which was a big challenge for all Chinese students.

在本句中，按照翻译界惯例，构成"which"引导的定语从句的字眼超过了8个，译为汉语单句，不可译作前置定语，以修饰、限定先行词"examination"，否则译文读来很累赘。因此，本句可译为："上高中时，我学习仅是为了高考。对所有中国学生来说，高考是一大挑战。"

E. g. I had read a lot of literature before I decided to explore the problem of farmers' income which was related to farmers' daily life.

在本句中，"which"引导的定语从句字眼没有超过8个，译作前置定语，修饰"income"即可。

本句中的"before"一词清晰地标明其前后两个主谓结构之间的先后关系。值得注意的是，翻译带有"before"引导时间状语从句的句子时，要合理安排主谓结构所表达的信息之间的逻辑关系。按照字面意思和汉语状语先行的原则，应先翻译"before"引导的时间状语从句，译为"在……前"，发生什么事。然而，地道的汉语经常不这样表意。译为汉语时，至少有2种处理方式：①将主句和"before"引导的时间状语从句分别表达的信息打个颠倒，即将主句信息译为"在……后"，将"before"引导的时间状语从句信息译作主句；②将整句信息按照"首先……（原英文主句信息），然后……"。笔者在此采用第二种处理方法，将本句译为："我首先阅读了大量文献，才决定研究与农民日常生活相关的农民收入问题。"

E. g. I love the atmosphere in school where everything makes me feel comfortable that any other place can't.

在本句中，"where"引导的定语从句超过8个，译作汉语单句，译文随即读来流畅。本句可译为："我喜欢学校的学习氛围。在这里，一切都让我感觉到了在其他任何地方感觉不到的舒适感。"

【参考译文】

个人陈述

直到上了大学我才真的了解自己的兴趣所在。上高中时，我学习仅是为了高考。对所有中国学生来说，高考是一大挑战。我大学主修经济学专业纯属偶然，然而这或许是迄今为止我做出的最明智的选择之一。我逐渐喜欢上了经济学，坚信研究经济学将成为我毕生的事业。

坦率地说，刚开始我并不知道经济学是什么。当时，我仅了解该领域内有诸多的经济学家及其不同的理论。我甚至曾担心学习经济学是否有用。直到有一天，我的经济统计学齐老师让我们利用学到的知识解决一个实际的经济问题。这是我第一次知道如何运用已掌握的经济学知识，并深切体会到只有掌握了一系列理论后才可能发现有意义的问题。我首先阅读了大量文献，才决定研究与农民日常生活相关的农民收入问题。然后我开始寻找数据处理工具,通过多次尝试我才找到了实用的工具,最终发现了可能影响农民收入的多个因素。当了解到我的研究结果与著名经济学家的意见一致时,我知道我先前付出的努力是值得的。之后,我意识到自己所学的经济学是能够解决人类生活实际问题的最伟大的科学。只要付出努力,就能不断进步。

如今,以经济方式进行思考已经成为我日常生活的一部分。例如,油价即将上涨对于赶在油价上涨之前去加油是个好机会,但我会思考油价上涨会对国际收支平衡造成的影响,以及政府应采取的货币政策等问题。因此,经济学已成为我生活中不可或缺的一部分。

别人认为简单的事情,我却觉得有趣而充满好奇,会花好几个小时加以思考。我会思考"被子晒干后有什么味道?""拉链的原理是什么?"等问题。您无法想象到每件小事可能都会引起我的兴趣。我会孜孜不倦地努力,直到找到答案。我曾经连续12个小时学习了解用于分析收集数据的工具(eviews)。经济学对我正如水对鱼一样重要。

我喜欢学校的学习氛围。在这里,一切都让我感觉到了在其他任何地方感觉不到的舒适感。我想成为一名教师,以便可以永远留在我已奋斗12年的这个学校。现在正是决定我未来人生走向的关键时刻。如果条件允许,我渴望接受获得经济学学位带来的挑战。

18.3.2.2　汉译英

个人陈述

我是2006年9月考入重庆大学数学与统计学院应用物理系,专业课是微电子和光电子。经过两年的学习后,对本专业的知识背景,有了深刻的了解。通过对信息光学处理实验的学习,发现自己对光学的相关知识很感兴趣。希望在研究生阶段可以利用光电子及纳米技术的相关知识,深入学习其相关器件的制备。

本科主要课程包括量子力学、固体物理以及数字电子技术和模拟电子技术等课程。本科三年,秉着严谨求学的态度,严格要求自己。在各种实验课上,完成每个实验。实验课学期结束的成绩都是优秀。在成绩方面,每学期均获得学院综合奖学金。分别获得甲等、乙等和丙等奖学金各两次,并且获国家奖学金一次。通过了国家计算机二级等级考试(C语言)。

同时,我也注意锻炼自己各方面的能力。我先后担任过班级团支书和院学习部副部长。可以很好的协同同学做好管理工作,曾被评为校优秀团员、优秀学生干部。2006年参加学校社会实践活动,获得二等奖。

我深知本科的学习,不足以满足以后的科研学习工作,因此,希望自己可以进入贵校进一步学习光电子及纳米技术等相关知识。研究生阶段的学习不同于本科学习。研究生阶段更注重动手能力的训练,以及自己快速获取提炼知识的自学能力,需要进一步钻研所学知识。

在研究生阶段,我希望做到三个方面的进步。

首先,是针对自己的英语知识的缺乏,注重自学提高运用英语的能力。通过暑假期间对雅思考试的准备,我相信英语是可以通过勤奋学习弥补的。

其次,在研究生期间,注重和老师交流沟通,培养自己快速获取知识,并且运用到实际实验中的能力。

最后,希望通过跟导师的学习,进一步掌握半导体光电纳米材料器件的制备方法,做到深入的学习。

在研究生学习期间,我会针对自己的本科学习经验,取长补短把更好的方法运用到学习实践中去。另外,在研究生学习结束后,会根据自己的学术能力,为自己制定更深入的计划,以做到最大限度的发掘自己的潜能,更好地服务社会,实现自身的价值。真心地希望各位老师,可以给我进入贵校继续深造的机会!

(https://wenku.baidu.com/view/6364240cba1aa8114431d9b4.html?from=search###)

1. 词汇层

应用物理系 Department of Applied Physics;微电子 microelectronics;光电子 photoelectronics;信息光学 information optics;纳米技术 nanotechnology;量子力学 Quantum Mechanics;固体物理 Solid-state Physics;数字电子技术 Digital Electronics Technology;模拟电子技术 Analog Electronics Technology;综合奖学金 comprehensive scholarship;班级团支书 league branch secretary of the class;优秀学生干部 "outstanding" student cadre;动手能力 hands-on ability;获取提炼知识 acquire and refine knowledge;雅思考试 the IELTS;半导体光电纳米材料器件 semiconductor photoelectronic nanomaterials and devices;发掘潜能 expand one's potential。

2. 句子层

这篇个人陈述若要翻译好并不容易,除了有一些专业术语外,更大的困难来自一些地道汉语表达法的英译上。同样,翻译时须注意准确措辞,合理搭建目的语句式,以最清晰的方式明示信息关系。

E.g. 我是2006年9月考入重庆大学数学与统计学院应用物理系,专业课是微电子和光电子。

"考入"应译作"be admitted to";翻译"重庆大学数学与统计学院应用物理系"需注意信息顺序,汉语概念从大到小,英语从小到大;"专业课是……"疑似措辞有误,应为"专业是……",若要与之前信息译作一体,还需注意形式上的转换。本句可译为:"In September 2006, I was admitted to Department of Applied Physics, College of Mathematics and Statistics, Chongqing University, as a microelectronics and photoelectronics major."

E.g. 本科三年,秉着严谨求学的态度,严格要求自己。

翻译本句时,译者须正确翻译"秉承严谨求学的态度,严格要求自己"。本句可译为:"During three years of undergraduate study, I maintained a rigorous approach to learning and made strict demands on myself."

E.g. 分别获得甲等、乙等和丙等奖学金各两次,并且获国家奖学金一次。通过了国家计算机二级等级考试(C语言)。

翻译本句时,译者须正确翻译"甲等、乙等和丙等奖学金",而且要有写长句的意识。"甲等奖学金"可译作"A-level scholarship",若照此翻译"乙等和丙等奖学金",就会重复使用"level"。为了避免重复,可以合并翻译,先总后分,将"甲等、乙等和丙等奖学金"译作"3 levels of scholarship, namely, A, B & C, ..."。本例中,逗号前后信息可用"and"连接,将英语译文拉长。故此,本句可译为:"I also won 3 levels of scholarship, namely, A, B & C, twice respectively, the National Scholarship once, and passed the National Computer Rank Examination Ⅱ (C language)."

E. g. 研究生阶段更注重动手能力的训练,以及自己快速获取提炼知识的自学能力,需要进一步钻研所学知识。

原文是汉语无主句。在本句所在的上下文中,"更注重……""获取提炼知识……"和"钻研所学知识"的逻辑主语应是"学生",为搭建译文句式,翻译时应该添加主语"the student"(表类别)。如果仔细阅读揣摩,我们发现,"以及自己快速获取提炼知识的自学能力,需要进一步钻研所学知识"若与前文译为一句英文,需挖掘、提炼出核心信息"能力",具体有二:①"快速获取提炼知识";②"进一步钻研所学知识",分别处理为动词不定式短语,作后置定语,共同修饰"能力"(ability)即可。故此,本句可译为:"On the postgraduate level, the student must pay more attention to developing hands-on ability and, through self-study, the ability to quickly acquire and refine knowledge as well as to probe deeper into what he/she has learned."

【参考译文】

Personal Statement

In September 2006, I was admitted to Department of Applied Physics, College of Mathematics and Statistics, Chongqing University, as a microelectronics and photoelectronics major. After two years of study, I had acquired a deep understanding and the background knowledge required for this major. Through the study of information optics and conducting experiments, I found that I was very interested in optics related knowledge. I hope that at the post-graduate stage I will be able to use my knowledge of photoelectronics and nanotechnology to prepare an in-depth study of related devices.

The main undergraduate courses include Quantum Mechanics, Solid-state Physics, Digital Electronics Technology and Analog Electronics Technology. During three years of undergraduate study, I maintained a rigorous approach to learning and made strict demands on myself. In all my lab classes, I completed each experiment carefully and conscientiously. My academic achievement in lab class was (rated as) outstanding at the end of a semester. With respect to academic achievement, I won the college comprehensive scholarship every semester. I also won 3 levels of scholarship, namely, A, B & C, twice respectively, the National Scholarship once, and passed the National Computer Rank Examination Ⅱ (C language).

Meanwhile, I also attached importance to developing all aspects of my abilities. I served successively as league branch secretary of the class and vice director of the Study Department. Moreover, I cooperated well with classmates in management work and was rated "excellent" school league member and "outstanding" student cadre. I participated in school activities — social and practical — in 2006 and won second prize.

I know that the undergraduate study is not adequate enough to meet the scientific research work in the future, so I hope to acquire further knowledge relating to photoelectron and nanotechnology at your institution. Study on the postgraduate level is different from undergraduate study. On the postgraduate level, the student must pay more attention to developing hands-on ability and, through self-study, the ability to quickly acquire and refine knowledge as well as to probe deeper into what he/she has learned.

At the postgraduate stage, I hope to make progress in three areas.

First, because my English is weak, I will focus on improving my ability to apply English through self-study. I believe that by studying hard for the IELTS during the summer holiday my English can be improved.

Second, during my postgraduate studies, I will emphasize interacting and communicating with teachers and will cultivate an ability to acquire knowledge quickly and apply it to practical experiments.

Finally, while studying under the guidance of an academic advisor, I hope to further master preparation methods for semiconductor photoelectronic nanomaterials and devices as well as to conduct an in-depth study.

During my postgraduate study, I will develop my strengths, based on my experience of undergraduate studies, so that I can apply better methods to my practice work. In addition, after concluding postgraduate study, I will make, according to my own academic ability, a more detailed plan than ever before and thereby expand my potential to the greatest extent while better serving society and realizing my own value. I sincerely hope that all teachers can provide me with the opportunity to continue my studies at your school.

练 习 题

一、请将下列文本译为汉语。

Personal Statement
By ×××

As a student of Birmingham University, I would like to scale higher intellectual heights by undertaking advanced studies in your Economics PhD program. When introduced to economics in college I realized that it interestingly qualified as a subject of both Arts and Science. It was an area defined by precise rules, principles and axioms and yet there was tremendous scope for self-expression in the form of interpretation and analysis. This facet of economics intrigued me very much and I decided to pursue further studies in Economics. But, at the same time, I realize that what I have learned as an MSc student is far from enough. Therefore, it is my desire to pursue a PhD degree in Economics at University of Birmingham.

Although I grew up as a general kid, I have to say I have a special childhood. Unlike most other Chinese students, I've started to study at home since I was almost seven years old. I never went to school till my parents was told I could not attend College Entrance Examination unless I graduated from high school. Self-study made me know how to deal with varieties of difficulties by myself and get the capability of self-control.

During the period of my previous study, I obtained a solid academic background in mathematics, computer science, economic theory and game theory. I have learned some basic economics theory when I was doing my bachelor degree. After I went to University of Birmingham I was fortunate to be able to take part in several great courses by excellent professors such as Macroeconomics, Microeconomics and Econometrics. These lectures not only introduced me to classic economics concepts and theories, but also broadened my perspective and gave me new insights into the depth of my field.

When I was an undergraduate student, I took a course named operational research which made me very interested in game theory. I found game theory so fascinating that I didn't hesitate to choose it when I saw it in my optional course list. Even though it is not easy I enjoy thinking and solving tough game theory questions.

Game theory is an area I would really like to explore. I am fascinated by game theoretic modelling of issues pertaining to Chinese traditional military treatise— The Art of War. I believe that game theoretic models can be effectively used in analyzing characteristics of wars and help us understand this book better.

Having experienced all the twists and turns in academic research, I have acquired the ability to conduct research independently. To make myself more insightful, I now need training in a broader theoretical frame work, which is why I am so eager to pursue further studies in your program. It is my hope that, upon the completion of my studies with you, I will be able to contribute more to the discourse on those topics.

(https://wenku.baidu.com/view/72f9dd9683d049649b6658fe.html? from=search)

二、请将下列文本译为英语。

个人陈述

尊敬的老师：

您好！我叫刘进顺，生于1996年10月庄浪县柳梁乡下岔村一个贫穷的农民家庭。虽然自幼家境贫寒，但我勤奋好学，励志要走出这个落后的村庄，走向文明发达的城市世界。

2012年9月我以优异的成绩考入庄浪县紫荆中学，这是我们县第二所重点高中学校。在校学习期间我依旧奋发图强，积极上进，在学习上一直是年级优秀学生里的先锋；在实践活动上，我担任了班长一职；在高中的三年里，我荣获了校级优秀学生的称号，优秀班干部的荣誉；也获得平凉市教育局颁发的平凉市自我飞翔励志成长优秀学生的荣誉证书；庄浪县县委宣传部颁发的庄浪县第四届校园文明之星荣誉证书。这些经历使优秀变成了一种习惯陪伴着我成长成才，使我有一种集体意识和为集体服务的责任意识，也锻炼了我组织校园活动的能力。

关于我的个人特长，我很诚恳地说并没有什么特别突出的地方，只是有一点我非常肯定，我喜欢读书，钻研科学问题，我觉得读书是一种快乐，它不仅激励着我不断上进，也丰富了我的精神世界；而对科学问题的思考和解决则是一种刺激，我喜欢这种刺激。因此我对自己将来从事的职业有了这样的设想:我希望能够在关乎国家前途与命运的重点行业从事科研工作,这不仅是为我的兴趣爱好,也是因为我有一种国家意识,想用我毕生的努力为国家的发展做出贡献。

  而贵校所设的专业正符合我的追求,因此我强烈希望能够成为其中的一员,并通过我的努力成为一个优秀的学子。为我国军事的强大和国防的安全,做出我的贡献!

  非常感谢您在百忙之中能够看我的个人陈述。

  此致

敬礼

自荐人:刘进顺
2015 年 4 月 22 日

(https://wenku.baidu.com/view/97fc7cb2da38376bae1fae1e.html? from=search)

各章练习题参考答案

第 1 章

一、请将下列文本译为汉语。

【参考译文】

学习推荐信

尊敬的先生/女士：

韩佳丽（Furnival Han）女士在学习英国文学过程中表现出的热忱和认真态度给我留下了深刻印象。能为韩女士写这封推荐信，我感到非常荣幸。作为她的老师和外国语学院主管，我非常了解韩女士对英国文学从厌恶、漠视到热爱这一态度转变过程。

韩女士在数学学科极有天赋，在大学入学考试中数学考了第一名。我记得刚进入大学时她对经济学感兴趣。她对英国文学由于不甚了解，起初很排斥，但经过一学期的学习，很快对该专业表现出浓厚的兴趣。虽然只是大一新生，她很快深入了解了相应的专业知识，成为我多年教学工作中见过的优秀本科生之一。

大二学习期间，在我们英语文学系留学归来老师指导和鼓励下，韩女士确定了未来前往美国继续深造的目标，为此付出不懈努力。她曾参加过多个文学专业论坛，甚至曾旁听硕士研究生课程。除认真阅读参考书外，她还大量阅读了西方专业书籍来扩充知识，这一点可从她的论文和日常交流中明显能感觉到。

此外，韩女士还是我们大学广播电台的优秀播音员和记者。她主持的节目曾受本校学生广泛喜爱。我为她的表现感到自豪，为她的文学素养感到惊叹。根据其专业优势、个人特点以及中国未来的发展趋势，我完全支持韩女士选择教育学专业。我非常希望像韩女士这样优秀的学生可以接受更好的教育，以加快中国教师队伍建设发展。

作为一名教授，我非常欣赏像韩女士这样有活力和好奇心、既勤奋又聪明的学生。当她告诉我她希望出国留学，我表示全力支持，鼓励她为此努力。作为学院主管，我自豪且衷心地向您推荐韩女士，我深信她将会给贵校带来全新的研究氛围。我衷心希望您能对她的入学申请提供帮助。

二、请将下列文本译为英语。

【参考译文】

Sales Contact Letter
Commodity Sales Arrangement

MM/DD/YY

Mr. ×××,

  Your last shipment of goods sold very well and was favorably received by our customers. Besides the quantity ordered in the telegram we sent you, I wish to request that you send the following samples since company staff intends to expand our scope of business:

  1. 2 suits of clothes for a 6-year-old boy

  2. 2 suits of clothes for a 6-year-old girl

  3. 2 suits of clothes (including skirts) for a 5-year-old child

  Please attach the relevant data and descriptions needed for reference, publicity and attending to consumer feedback. I expect that your products will be popular. Once our company has made its decision, we will order from you on a large scale. We will contact you later as to the specifics. Please send your samples as soon as possible. Thank you for attending to this matter. I am looking forward to your reply. Please enclose the price and a quality specification description of the samples with the data.

第 2 章

一、请将下列文本译为汉语。

【参考译文】

前　　言

  在成千上万种计算机滥用方式中,模拟位居榜首。造成计算机大量滥用的原因不难发现:

  1. 每种模拟都在模仿某种事物,但是计算机应该模拟仿真人员脑海中的想法却没有什么特殊原因。

  2. 尤其对于那些在黑暗中摸索前行并迫切找到指路明灯的人而言,他们倾向于把计算机输出结果误认为是绝对真理。

  3. 模拟语言已经可以更容易地实现令人印象深刻的模拟,但要更容易地获得有效模拟却并不容易。

  4. 当前无任何基于广泛实践经验开设的课程。因此,任何人,不管用何种语言编写模拟程序,产生了何种结果,编写完后均是"专家"。

  5. 模拟发展前景很大,很容易将希望和成果混为一谈。

  存在以上原因皆因缺乏知识。一本好书可以帮助人们弥补所有不足吗?不幸的是,答案是"不能",而且永远是"不能"。这是个非常大的问题。书籍本身从不会发出任何行为。

尽管好书可为"人们"的活动提供帮助，但是不管书籍有多好，书籍本身只能被动地提供信息。

《模拟：原理和方法》是一本非常值得一读的好书。如果使用恰当，作为系统思考的工具，该书可以帮助我们恢复对模拟的信心。因此，我们将该书选入计算机系列丛书中。尽管如此，我觉得有必要说明"恰当使用"的含义。

与其他诸多模拟书籍的作者相比，该书的作者格雷比尔(Graybeal)和普齐(Pooch)写作的一大优点是保持客观性。直到现在，大多数重要的模拟书籍都是由一些特定语言或方法的提倡者撰写的。这类书的政治性和教育性很强，所以作为一门课程适合在初级阶段学习。但是，我们现在已经成熟，不再因不了解计算机 A 语言或 B 方法而使用 X 语言或 Y 方法进行模拟。通过学习格雷比尔和普齐编著的《模拟：原理和方法》，模拟学习者在选择问题解决可选方案时会获得更加均衡的信息，而不仅仅读了一篇关于"为什么我的方法是解决问题的最好方法和唯一应该知道的方法"话题的论文或专著。

当提到模拟"学习者"，我希望借此介绍本书的另一项重要特征——表达的清晰度。迄今为止，大多数人自学模拟知识。当遇到问题时，人们会咨询朋友和查阅有关模拟的书籍。因为要解决问题，他们阅读时非常有动力，即便是作者编著的内容非常晦涩难懂。

今天，"学生"听众越来越多。他们会走进正规课堂，在经验丰富老师的帮助下学习模拟技术。这些学习者需要不同类型的书籍，本书可以满足这种需求。本书表意清晰，但首次提到一个话题时不会提供超过学生所需的细节信息。此类例子比比皆是，相当有教学价值。本书按照学生能力发展渐进的方式，向学习者提供测试理解所需的练习。简言之，本书是"一本教科书，而不是模拟手册"，这才是我们现在想要的书籍。

我并不是指任何未经过课堂学习且初涉模拟人员不应使用本书。相反，我认为对于立志成为模拟专家的人员来说，应遏制阅读计算机语言手册类书籍的冲动，而应首先阅读本书，这样会获得更好的学习效果。本书表述清晰易懂，上进心强的读者不但可以从该书中学到很多课程，而且可以为阅读需要理解力更强的手册类书籍做好充足的准备。

尽管如此，我无法要求急切希望了解模拟知识的学习者慢下脚步，静心阅读必要的相关书籍。如果可能，或许我会通过立法禁止当前最糟糕的模拟滥用行为。由于没有立法权，我只能向读者推荐我们的系列好书（比如这本书），希冀广大教师将这些书当教科书用，这样读者就可以规避上演新的模拟悲剧。

二、请将下列文本译为英语。
【参考译文】

Network Technology

Network technology, a new technology dating from the mid-1990s, integrates scattered Internet resources into an organic whole, and enables overall sharing of resources as well as integrated cooperation, thus providing people with a an integrated capacity to use resources transparently and acquire information as required. These resources include high-performance computers, storage resources, data resources, information resources, knowledge resources, and expert resources as well as large-scale databases, networks, sensors, etc. Although the current Internet is restricted to information sharing, it is regarded as the third stage of Internet development. The network may be a regional network, the internal network of an enterprise or institution, a local area network (LAN), or even a home or personal network. Its basic feature is not necessarily its scale, but the sharing of resources and the elimination of isolated resource islands.

Key Technologies

The key technologies of the network include the network node, the broadband network system, the resource management and task scheduling tools, and the visualization tools of the application layer. The network nodes are the providers of network computing resources, including high-end servers, cluster systems, large-scale MPP system storage devices, databases, etc. The broadband network system is necessary for providing high performance communication in a network computing environment. Resource management and task scheduling tools are used to address key issues such as resource description, organization, and management. The task scheduling tool executes the dynamic scheduling of tasks in the existing system according to the load condition of the system, so as to improve the system's operational efficiency. Network computing is mainly scientific computing, often accompanied by huge amounts of data. Converting the results into information directly observable on a graph can help researchers free themselves of the difficulties involved in understanding data. This requires the development of visual tools that can be transmitted and read in the course of network computing while providing a user-friendly interface.

Existing Problems

In the previous chapter, we presented a brief history of the Internet and all aspects of its applications. Due to its rich and colorful contents, the Internet is attracting more and more users and is gradually permeating every aspect of our life and work, thus greatly transforming long existing traditional modes of thought and life styles. As for the Internet itself, active user participation is causing it to expand so rapidly on a global scale that its endurance capacity faces severe challenges, such as a shortage of bandwidth and IP address resources...

第 3 章

一、请将下列文本译为汉语。
【参考译文】

"神舟六号"宇宙飞船

"神舟六号"任务标志着中国开启载人航天实验的新征程

周日，一位航天高级工程师在此声称，中国计划发射第二艘载人航天宇宙飞船"神舟六号"，这标志中国的载人航天进入了真人参与的空间科学试验新阶段。中国载人航天工程总设计师王永志表示，太空是独特的、强辐射的、低重力的真空环境，"神舟六号"宇宙飞船发射后宇航员可以在太空进行科学实验。

2003年10月,"神舟五号"成功发射,中国成为全球第三个成功实现载人太空飞行的国家。在为期一天的太空飞行中,"神舟五号"载着唯一的宇航员杨利伟环绕地球飞行了14次,但杨利伟始终没有离开返回舱中的座椅。王永志指出,继成功完成"神舟六号"飞行之后,中国将实施宇航员太空行走、太空舱与空间模块对接、发射空间实验室等更为宏伟的计划。开展太空飞行工作是最复杂、最困难的航天工程,展示的是一个国家的科研能力和经济实力。王永志声称:"这是扩大人类生存空间、开发和利用太空资源的重大举措。中国永远不做超级大国,但作为拥有13亿人口的世界上最大的发展中国家,中国应该在全球航天事业发展中占有一席之地,做出应有的贡献。"

成功发射

周三上午,中国将载有费俊龙和聂海胜这两名宇航员的第二艘载人宇宙飞船"神舟六号"送入太空。这次太空飞行持续了数日。上午9时,位于中国西北部的酒泉卫星发射中心用"长征二号"F型运载火箭发射了"神舟六号"载人宇宙飞船。39分钟之后,酒泉卫星发射中心宣布发射成功。中国国家主席胡锦涛、总理温家宝等领导人亲临现场目睹了这次发射。据估计,大约数以亿计的中国人通过电视直播、广播和互联网了解这次发射的现场报道。"神舟六号"载人宇宙飞船的成功发射进一步证明了中国科技日新月异的进步。两年前,"神舟五号"载人宇宙飞船的成功发射标志着中国成为继前苏联和美国之后第三个实现载人航天飞行的国家。与2003年10月将宇航员杨利伟送入太空的首次载人飞行不同,这次航天任务考验了中国实施多人多天航天飞行的能力。此外,"神舟六号"航天任务还包括"真人参与并开展科学实验"活动,提高了太空实验水平。自发射以来,"神舟六号"飞船运行良好。当天下午3点50分,航天器由椭圆飞行轨道转到了圆形飞行轨道上。大约两小时后,宇航员从返回舱成功进入轨道舱进行科学实验。

二、请将下列文本译为英语。

【参考译文】

With the rapid development of urbanization in China, intensive, efficient, energy-saving and environmentally-friendly public transportation has become a strategic option for creating a resource-saving and environment-friendly society. Since 2011, China has been vigorously promoting an important initiative to create "public transit cities", and has encouraged the development of ground mass transit such as Bus Rapid Transit (BRT), with an emphasis on "improving the level of public transportation facilities and equipment". By using special separate lanes, a high-quality infrastructure, and offering rapid & frequent operations, BRT provides swift, comfortable and low-cost service. Unlike the simple booth at a conventional bus stop, a BRT station is a large complex which, in addition to being a good place for passengers to experience public transportation for themselves, is a relatively independent building facility for the urban road system. The operating efficiency of the system is affected by the form and layout of station, while its attractiveness is a function of the comfort provided by the station environment. For this reason, the structural design of BRT stations is one of the decisive factors determining the system's quality. According to statistics, BRT has been implemented in more than 160 cities around the world, including nearly 30 cities in China. However, the existing literature is short on research regarding the architecture of BRT stations, especially with respect to thermal comfort and energy saving. This paper attempts to explore the main strategies used in BRT station architectural design by analyzing some existing cases.

1. Features of BRT Station Buildings

First, the composition of a BRT station is relatively complicated. Generally speaking, a BRT station has an equipment room, a ticket booth, a check-in gate, and waiting areas. In order to reduce the boarding time, BRT uses the subway system's pre-boarding ticket-checking method, which requires a separate waiting area.

Second, the form, location and size of a station depends primarily on the system's operational mode, passenger flow demand, the number of lines, and road space. The station should be designed to facilitate system operation and management.

Third, a station is built on a large scale, usually exceeding 30 meters in length. Moreover, the boarding area will increase with the growth of the number of substations and the demand for larger spaces to park long buses. The width of the station, usually exceeding 3 meters, depends on its structure and passenger flow volume, while the height of the platform is the same as that of the vehicle floor to allow convenient horizontal boarding.

Fourth, the internal environment of the station is greatly affected by the surrounding streettraffic since all special BRT corridors are located in the middle of the road, in order to improve vehicle operation efficiency. Because of the high temperature of the asphalt pavement, the serious pollution produced by automobile exhaust fumes and the lack of green shade trees in the surrounding area, the comfort level of the station environment can be rather poor. Therefore, a station must be provided with facilities for climate control in order to make the waiting environment more comfortable.

Fifth, compared with other mass-transit, BRT has the advantage in terms of low production costs and the short time required for implementation. Accordingly, BRT stations aim at both low production costs, easy construction, and standardization. Primary building materials include such common materials as steel, concrete, and glass, which are safe, durable and easy to assemble. However, these materials tend to absorb a large amount of thermal radiation, thus increasing the temperature inside the station. Moreover, in order to reduce cost and energy consumption, climate-regulating mechanical equipment like air-conditioners and heaters are rarely used. Rather, station design emphasizes the selection of architectural forms and components as a means to "passively" adjust the thermal environment in a station…

第 4 章

一、请将下列文本译为汉语。

【参考译文】

本论文利用从英国国家语料库(BNC)和西班牙皇家语言学院数据库(CREA)中收集的跨语言实证数据,根据符号学的文化概念,综述出黑白两色隐喻概念。上述两个用于比较的语料库均包含字面含义和比喻含义的词汇,但我们关注的重点仅仅是比喻义,将多词单元(如词汇搭配、成语或谚语等)与其使用的不同文化背景联系起来确定词义。

本论文的研究问题是:在英语和西班牙语中,文化实际上是如何在与"黑""白"相关的隐喻和转喻表达法中表现出来的?表现程度如何?为寻求这一问题的答案,我们使用Piirainen的分类法(Piirainen E,《从文化符号学角度了解短语》,布尔赫尔等人编著,第209页至第219页,2007年出版),以便从定性视角分析不同的文化现象。研究结果表明,文化隐喻似乎需要了解其输入域和属性,或输入域和输出域之间的关系。

通过比较包含"黑色/negro"和"白色/blanco"的短语,明显可以看出短语学的文化基础(沃尔兹博克,《语义学精华和一般概念》,牛津:牛津大学出版社,1996年出版)。使用"黑色/negro"表达"不幸或不高兴"意思和使用"白色/blanco"表达"好或无辜"意思均反映文化事实。如果使用此类短语表述物理实体(颜色术语),则可表示物体的这些属性。英语和西班牙语中与"黑色"和"白色"搭配的词语、成语和谚语都是有力的文化符号。语言使用者对语言和文化中"黑色"和"白色"符号之间的关系了解得越透彻,在不同语境中阅读理解能力就越强。

二、请将下列文本译为英语。

【参考译文】

Letter of Apology

×××ingda Trading Co., Ltd. in ××City,

We have received your letter sent on ××, 20××. In response to your complaint, we apologize for the fact that some of the 35 sets of Huanghua computer desks in the Computer Desk Purchasing Contract agreement signed on Jan. 7, 20×× have interface fractures. Our party attaches great importance to this matter and has carried out an investigation.

Investigation of relevant departments: The ××× type Huanghua computer desks produced by our factory all pass through quality control and are high-quality products. The interface fracture in some of the computer desks mentioned in your letter was caused by improper handling by our workers when the desks were taken out of storage. We apologize to your company again for your losses. Please advise as to the number of damaged computer desks and the degree of damage sustained as soon as possible and forward the notarial and inspection certificates, and our company will provide compensation unconditionally according to the actual loss as soon as possible.

We will take this as a warning, and, in the future, check for problems and shortcomings in our work while taking corrective measures to put an end to such occurrences.

We hope for your understanding in this matter and that we will be able to continue our excellent business relationship in the future.

Awaiting your reply!

Guangming Furniture Co., Ltd. in ×××City

第5章

一、请将下列文本译为汉语。

【参考译文】

五年前,迈克尔·普雷斯曼(Michael Preysman)曾发誓,称他的在线服装公司艾芙兰(Everlane)永远不会开设实体店。

普雷斯曼曾在接受《纽约时报》采访时声称:"我们在进入实体零售领域前将关闭公司。"三月份,普雷斯曼接受 Quartz 采访时再次重申,"没有一家服装店可提供'最好的客户体验'"。

但现在看来,就是否开设实体店普雷斯曼的看法已发生改变。

六年前,普雷斯曼创立心系社会的品牌艾芙兰。12月2日,该品牌名下的两家实体店将正式营业。这两家旗舰店位于纽约的王子大街和旧金山的教会区。艾芙兰将在这两家实体店销售该公司的许多最畅销产品,如T恤衫、羊绒衫、牛仔裤和鞋子等。尽管首批实体店的店面面积都相对较小〔纽约的旗舰店占地仅2 000平方英尺(约186平方米),旧金山的旗舰店占地3 000平方英尺(约279平方米)〕,但普雷斯曼声称,两家实体店开门营业可为公司吸纳新客户,也可与现有顾客开展更为密切的互动。

普雷斯曼声称:"客户总是告诉我们,在购买产品之前他们想接触产品。我们认识到,如果我们希望成为全国乃至全球性的大企业,就需要建立自己的实体店。"

普雷斯曼说,当然还有其他一些原因。例如,他们发现,即便是喜欢网购的客户更愿意亲自退货或换货。

普雷斯曼创立艾芙兰时年仅25岁。该公司是最新开设实体店的在线零售宠儿。2013年,最受欢迎的在线眼镜零售商瓦尔比·派克公司(Warby Parker)在纽约开设第一家零售店。如今,在美国和加拿大该公司已拥有60余家实体店。近年来,男装零售商Bonobos也在实施线下实体店扩张策略,其旗下已创立M. Gemi鞋业公司和Cuyana服装品牌。今年,沃尔玛以3.1亿美元收购了Bonobos公司。

这一发展趋势突显零售商对技术态度的快速转变。不久前,一些零售商还认为开实体店是过时的做法,只会增加营业费用,但事实证明,实体店在吸引新顾客和完成在线订单方面同样具有重要作用。今年,为进入实体店业务领域,美国电商巨头亚马逊以137亿美元的价格收购了全食超市(亚马逊首席执行官杰弗里·贝佐斯拥有畅销报业公司《华盛顿邮报》)。

市场研究企业弗瑞斯特公司零售业分析师斯科瑞塔·姆普鲁（Sucharita Mulpuru）声称："在您摘到低处的果子之后（客户愿意在线购买您的品牌或愿意在社交媒体上关注您的品牌），您还有其他什么办法让顾客购买您的产品吗？在此情况下，实体店可以彰显其影响力。"

姆普鲁补充道，由于购物中心存在大量虚席以待的店铺，开设小型实体店的市场条件也有所改善。鉴于全球范围内普遍出现商店关门潮（迄今为止，今年已有7 000多家商店关闭），在月租或年租方面业主和购物中心也愿意和小众品牌公司签订灵活的协议。

二、请将下列文本译为英语。

【参考译文】

Taiwan media reported that when the Chinese Liaoning aircraft carrier formation crossed the Miyako Strait on the 25th, a Japanese submarine attempted to sneak into the formation at which point the aircraft carrier frigate immediately dispatched Zhi-9 anti-submarine reconnaissance helicopter. Tracking the submarine with sonar, the helicopter advanced to within 10 kilometers of Miyako Island.

Nippon Television reported, quoting information from the Joint Staff Council, that a Chinese Zhi-9 helicopter took off from the aircraft carrier frigate, crossed the Miyako Strait and flew in a southwest direction into airspace 10km–30km distant from Miyako Island; it did not, however, violate Japan's air space. The Japanese Air Self-Defense Force immediately dispatched a fighter plane in response.

Taiwan media believed that the Japanese submarine may have been detected and intercepted by the Zhi-9 helicopter.

The Japanese Ministry of Defense said that a Samidare frigate and a P3C anti-submarine alert aircraft were dispatched from the Naha Military Base to surveil and track the Liaoning aircraft carrier formation, but didn't give any news regarding the surveillance and tracking of the Japanese submarine.

第6章

一、请将下列文本译为汉语。

【参考译文】

中国国务院总理温家宝在酒泉卫星发射中心发表了简短的重要讲话。温总理高度赞扬神舟六号的成功发射，称这将载入中国的光荣史册。温总理指出，中国发展太空计划完全出于和平目的，是对人类科学与和平事业做出的贡献。他补充道，中国愿意与其他国家合作，推进太空科技向前发展。中国已经实施三级载人太空飞行计划，最终会建立一个永久性空间实验室。

"神舟六号"宇航员完成更多项太空测试

周四，参与中国第二次载人航天飞行的两名宇航员在"神舟六号"宇宙飞船的返回舱和轨道舱之间往返行走，继续开展相应的实验，为未来宇航员太空行走、对接演练和建立空间实验室等更为宏伟的计划做好准备。中国政府的官方媒体新华社报道,宇航员在返回舱和轨道舱之间往返行走主要是为了测试宇航员行走对宇宙飞船的影响。周三,费俊龙首次离开宇航员太空舱,进入轨道舱。周四,费俊龙的同事聂海胜利用安装在轨道舱上的设备完成了类似的工作。载有两名宇航员的太空飞行和宇航员进入轨道舱标志中国太空计划已取得新进展。

成功着陆

周一早些时候,中国第二艘载人宇宙飞船"神舟六号"返回舱在内蒙古偏远地区安全降落。"神舟六号"宇宙飞船成功完成了连续五天的飞行任务,推动中国宏伟的太空计划向前发展。当天上午4:32(格林尼治标准时间周日20:32),"神舟六号"返回舱成功着陆。在经历了115.5小时的飞行之后,宇航员费俊龙和聂海胜出舱时受到英雄般的欢迎。中国共产党领导人吴邦国表示,"成功完成这次航天任务对提升中国在世界的威望、促进中国经济、科技和国防的发展和民族凝聚力具有重要意义。"中国载人航天工程办公室主任唐贤明在接受记者采访时称,中国政府在"神舟六号"载人航天飞行总共花费了9亿元人民币(折合1.1亿美元)。负责"神舟六号"载人航天飞行的最高军事指挥官陈炳德宣布,"'神舟六号'载人航天飞行取得圆满成功"。费俊龙和聂海胜乘机抵达北京之后,受到陈炳德和中国国防部长曹刚川率领的军方领导人的热烈欢迎。两位宇航员与家人团聚。根据中国政府的官方媒体新华社报道,费俊龙和聂海胜声称"身体状态良好"。两位都是中国人民解放军上校,都曾是战斗机飞行员。返回舱着陆后,费俊龙、聂海胜相继走出返回舱。最初的医疗检查结果表明,两名宇航员"身体体征正常"。返回舱着陆时间比原计划提前了约3个小时。返回之前,两位宇航员将返回舱和轨道舱分离。轨道舱处于宇宙飞船的"鼻部",将作为电子和光学监视卫星继续为期8个月的在轨飞行。

二、请将下列文本译为英语。

【参考译文】

Features

Computer systems are characterized by their capacity to carry out high-speed, accurate calculations and judgments as well as by their universality, ease of use, and connectability to a network. ①Calculation: Almost all complicated calculations can be realized by means of arithmetical and logical operations. ② Judgment: A computer has the ability to distinguish between different situations and select different processes. Accordingly, it can be used in the fields of management, control, confrontation, decision-making, reasoning and so on. ③ Storage: A computer is able to store huge amounts of information. ④Accuracy: As long as the word length is sufficient, calculation accuracy is, theoretically, not restricted. ⑤Fast speed: The time required for a computer to perform an operation is so short that it is measured in nanoseconds. ⑥ Universality: Since a computer is programmable, it can execute different programs that implement different applications. ⑦Ease of use: Rich high-performance software and intelligent man-machine interfaces greatly facilitate the use of computers. ⑧ Networking: Multiple computer systems are able to transcend geographic boundaries so as to share remote information and software resources with the help of communication networks.

Composition

Fig. 1 shows the hierarchy of a computer system, the core of which is the hardware system, the actual physical device for information processing. The outermost layer is the person who uses the computer, that is, the user. The interface between the user and the hardware system is the software system, which can be divided into three layers: system software, supporting software and application software.

Hardware

The hardware system is mainly composed of the CPU, the memory, the input and output control systems and various peripheral devices. Specifically, the CPU is the main component for processing information at a high speeds reaching up to several hundred million operations per second. The memory, which often consists of a fast central memory (with a capacity reaching hundreds of megabytes, even several gigabytes) and a massive slow-speed secondary storage memory (with a capacity of tens of gigabytes or more) is used to store programs, data, and files. Various input/output peripheral devices function as information converters between man and machine, while the input-output control system manages the information exchange between the peripheral devices and the main memory (CPU).

Software

There are three types of software: system software, supporting software and application software. System software is composed of an operating system, utility program, compiling program and so on, while the operating system implements the management control of various hardware and software resources. Utility programs are designed for users' convenience, such as in text editing. The function of a compiling program is to translate a program written by a user into assembly language or to translate some advanced language into a machine language program executable by a machine. The supporting software, which includes interface software, tool software, environment database, etc., supports the environment for computer use and provides tools for software development. The supporting software may also be regarded as part of the system software. Being the outermost layer of the software system, the application software, which runs with the help of system and supporting software, is a special program written by users according to their needs.

第 7 章

一、请将下列文本译为汉语。

【参考译文】

网络拓扑

网络物理布局的重要性通常低于网络连接节点的拓扑结构。用于描述物理网络的大多数图表均为拓扑图,而并非地理位置图。拓扑图上的符号通常表示网络链接和网络节点。

网络链接

　　用于链接设备构成计算机网络的传输介质(在文献中通常称为物理介质)包括电缆(以太网、家庭网络、电力线通信、G.hn等)、光纤(光纤通信)和无线电波(无线网络)。在 OSI 模型中,此类传输介质处于第 1 层和第 2 层,即物理层和数据链路层。

　　局域网(LAN)广泛采用的传输介质统称为以太网。利用美国电气与电子工程师协会(IEEE)802.3 文件明文界定的介质和协议标准,可使以太网中的网络设备相互通信。以太网通过铜缆和光缆传输数据。无线局域网标准(如 IEEE 802.11 明文定义的标准)使用无线电波或其他红外信号作为传输介质。电力线通信使用建筑物中的电力电缆传输数据。

有线技术

　　可从终端发射光线的玻璃线束。

　　光缆将数据从一台计算机/网络节点传输至另一台计算机/网络节点。

　　有线技术可按照传输速度从最慢到最快排序。

　　同轴电缆广泛用于有线电视系统、办公楼和其他采用局域网的工作场所。此类电缆由绝缘层(通常为具有高介电常数的柔性材料)包裹的铜线和铝线构成,绝缘层自身也被导电层包围。绝缘层有助于减少信号干扰和信号失真。同轴电缆的传输速度在 2 亿比特/秒至 5 亿比特/秒之间。

　　在家庭布线(同轴电缆、电话线和电力线)中,国际电信联盟电信标准化部(ITU-T)G.hn 技术通常用于创建高速局域网(传输速度最高可达 10 亿比特/秒)。

　　由两根铜线绞合制成的双绞线用于所有电信领域,是使用范围最广的介质。普通电话线由两根相互绝缘的铜线绞合制成。计算机网络电缆(IEEE 802.3 定义的以太网电缆)由用于传输语音和数据的四对铜线构成。使用两条绞合的导线有助于减少串扰和电磁感应现象。双绞线的传输速度在 200 万比特/秒至 100 亿比特/秒之间。双绞线有非屏蔽式双绞线(UTP)和屏蔽式双绞线(STP)两种形式。每种双绞线均可细分为不同的等级,便于在不同环境中使用。

　　光纤是玻璃纤维,用于承载表示数据的光脉冲。与金属导线相比,光纤具有传输损耗极低和不受电干扰影响等优点。光纤可以同时承载和传输多个不同波长的光线。光纤不仅可以大幅提高数据传输速率,还有助于实现每秒数万亿比特的数据传输速率。光纤可用作长距离高数率传输的电缆,也可作为连接大陆的海底电缆。

　　在商业活动中,价格是决定有线和无线技术选择方案的重要因素。利用无线技术选择方案提供的价格溢价,用户可购买有线计算机、打印机和其他设备,由此可为用户提供更好的价格优势。在做硬接线技术产品决策时,必须慎重审查可选技术的约束和限制条件。在特定环境下,商业需求和员工需求的优先级可能会高于成本考虑。

二、请将下列文本译为英语。

文本 1：

【参考译文】

Notice to Tourists

In order to create a beautiful, orderly and clean environment in this park, where tourists can enjoy high-quality services while taking a walk, participating in leisure and recreational activities or simply strolling about for fun, please abide by the following regulations：

1. All tourists should strictly observe the "Seven Don'ts" appearing in the Notice issued by the municipal government.

2. Bicycles and motor vehicles are prohibited in the park.

3. No pets are allowed in the park.

4. We request that all tourists obey the management, pay attention to hygiene, and observe social norms.

5. Bird catching is forbidden.

文本 2：

【参考译文】

The Yangtze River originates on the "roof of the world", i. e., the southwest side ofGeladaindong Peak in the Tanggula Mountains on the Qinghai-Tibet Plateau. The river's stem streams flow through 8 provinces, namely, Qinghai, Sichuan, Yunnan, Hubei, Hunan, Jiangxi, Anhui and Jiangsu, 1 autonomous region, i. e. Tibet and the 2 independent municipalities of Chongqing and Shanghai, before they pour into the East China Sea, east of Chongming Island. The overall length of the Yangtze River is about 6,300 kilometers, which is longer than the Yellow River (5,464 km) by more than 800 kilometers. In terms of length, the Yangtze River, which is located exclusively in China, ranks third in the world, falling behind the Nile in Africa and the Amazon River in South America. The former, the longest river in the world, flows across nine African countries and the latter, the second longest on earth, across seven South American countries.

第 8 章

一、请将下列文本译为汉语。

【参考译文】

冬日度假地：读者旅游指南

获胜技巧：前往牙买加安东尼奥港附近村落

如果您梦想在沙滩上享受30℃的阳光，休闲地度过这个冬天，但您的资金只适合在士嘉堡度过周末，为何不考虑前往梦想之地附近的村庄呢？在牙买加安东尼奥港附近蓝山山脚下的村庄中,当地居民向游客提供非常优惠的"木瓜"木屋服务。您不仅可享受膳宿服务和迤逦的自然风光,还可以悠闲地躺在阳台吊床上读书。在这里,您只会听到啄木鸟和鹦鹉的鸣叫声以及树叶的婆娑声。这里没有旅游业,不会有人劝您购物,只有当地居民迎接每一天的相互问候和灿烂笑容。"木瓜"木屋包含两间卧室,可供四个人休息。一月来此地旅游,您每晚仅需支付10美元。无须惊叹,赶快收拾行囊出发吧!

马丁·柯莱特

冈比亚库图市

(图)

摄影:阿拉米

在洒满阳光的海滩上轻啜冰啤是我理想的冬季假期度假方式。与其他诸多冬日度假胜地相比,令人放松且风景优美的库图市冈比亚海滩不仅费用更低,而且还有很多酒吧和饭店。这里的库姆博海滩酒店(Kombo Beach Hotel)内设多种风格的酒吧和饭店,氛围极为悠闲。在十二月份,每天两人住宿(含早餐)费用仅约70英镑。这个酒店还设有当地最受人欢迎的利瓦(Liv)酒吧,这样您可以省下前往附近别处该酒吧的返程出租车费。

葡萄牙阿尔沃市

如果想在冬天逃离英国,前往自带私人屋顶平台、设备齐全的公寓,如果从公寓走几分钟即可到达自然淳朴且摩尔人文化浓郁的阿沃尔(Alvor)渔村和美丽的沙滩—您是否会心动呢?在业主自营(OwnersDirect)平台上,我们每晚只需支付35英镑的住宿费用。在这里,我们和当地人一起购物,一起享用美食(通常在每天早晨),真正体验了葡萄牙社区生活。在慵懒的假日里,我们沿着木板路和河口步行道漫步,享受海滩美景。乘坐前往法罗市的廉价航班,使冬日逃离英国度假计划显得更加经济实惠。

古巴乌斯怀亚市

我们的住所是一家私营旅馆(通过 cuba-junky.com 订购,每晚住宿每个家庭仅需支付40英镑),位于古巴猪湾乌斯怀亚市风景如画的海滩上。在这里,您不仅可以看到悠闲漫步的火烈鸟,而且还可前往库埃瓦德洛斯山脚下的海中潜水。此外,这里还充满浓郁的历史和文化氛围。因美国入侵而英勇牺牲的古巴烈士安眠与此,墓碑上记载着他们的牺牲地。回哈瓦那途中,我们乘坐的出租车是20世纪50年代生产的别克车。为防止螃蟹破坏轮胎,当地使用美国黄色校车供去潜水的游客搭乘。

二、请将下列文本译为英语。

【参考译文】

A Summary of the CSRC press conference on August 21, 2015

20:04 August 21, 2015   source: China Economic Net

On August 21, 2015, a press conference was held by CSRC (China Securities Regulatory Commission). Zhang Xiaojun, the spokesman, informed the public of some illegal share-holding cases involving big shareholders and de facto controllers of listed companies which had recently been investigated. He then spoke in favor of related policies involved in opening up the expanded securities institutions in Hong Kong and Macao (see news columns on official websites). Finally, he answered reporters' questions.

Q: Have the supervisory departments suspended the approval work on structured funds? How is it going with the guiding policies involved?

A: In view of the complicated situation of the mechanism of structured funds, which average investors cannot easily understand, and of some new situations and new problems that have occurred before, the registration of such products will be deferred accordingly and policies involved are under study.

Q: When will the 2nd amendment to *the Securities Law* take place? How is it going now?

A: The revision on *the Securities Law* is a great event in the securities market, attracting the attention of people from all walks of life. In April of this year, the amendment to *the Securities Laws* was reviewed for the first time at the 14th conference of the standing committee of the twelfth National People's Congress. At present, the revised draft is being modified by the legislature according to specific legislative procedures. The CSRC will cooperate fully with the legislature concerned in carrying out the related work as well as studying and giving feedback on legislative requirements and suggestions in the hope that the second revision of *the Securities Law* can be arranged as soon as possible.

Editor in charge: Wei Jingting

第9章

一、请将下列文本译为汉语。

【参考译文】

"双速"互联网

网络中立的支持者主要关注套餐推行中的两个主要问题。问题一：如果将互联网服务定价分解为套餐定价，就会造成定价混乱，用户将难以比较不同的服务，从而掩盖了移动运营商和互联网服务提供商增加用户使用总成本并将利差收益纳入囊中的事实。

问题二(更具系统性)：获得优惠待遇的特定应用会把互联网用户划分为"富人"和"穷人"。有时被称作"双速互联网"的服务会增强处于行业顶端公司的地位，而创业公司发展因此达不到行业顶端公司的规模。

例如，如果一家公司试图向新西兰国内的沃达丰客户推广新视频流媒体服务，则该公司除了利用自身的竞争方式(可能是更好的节目或更便宜的订阅费用等)打败Netflix，还需要面对一个现实问题：部分沃达丰服务订购用户依然会定制Netflix服务，而不会超出套餐流量上限，但是创业公司使用流媒体服务则很快达到流量上限。

在全球其他区域，尽管网路中立性处于较弱状态，从未出现广泛的负面影响，为何美国如此强烈且消极地对待国内网络中立性的变化？让我们来看一下自称为互联网头版的红迪网站(Reddit)，了解用户对网络中立的反应：该网站主页的25个报道中有16个报道与网络中立相关。在该页面的所有链接中，除两个链接之外，其余链接均导航至同一网页。

在英国出生的旧金山风险资本家本尼迪克特·埃文斯(Benedict Evans)注意到，最大的

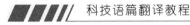

区别是竞争。埃文斯写道:"在伦敦居住时,我可以从数十家宽带服务商中选择服务。这使得网络中立成为更具理论性的话题。"

在美国,绝大多数居民在购买宽带服务时基本没有选择权。依法成立的本地垄断公司和全国双头垄断公司为四分之三的美国居民提供高速网络连接服务。这样,互联网服务提供商即拥有更大的权力。互联网服务提供商借助网络中立,要求客户支付额外费用才可全速浏览自己喜欢的网站,或迫使网站公司为获得客户而额外付费。

权力动态的变化

2007年11月,巴拉克·奥巴马(Barack Obama)在谷歌公司加利福尼亚州总部的市民会议上发表讲话。在这次访问中,奥巴马谈到他对网络中立的态度。(摄影:金伯利·怀特/盖蒂)

自2003年首次出现"网络中立"这一术语以来,有关网络中立的争斗已持续10余年,但近几年来,争斗格局已发生变化,主要是因为美国总统选举。新总统在制定决策过程中,似乎以取消前任总统的所有措施为主要目标。

与此同时,大型互联网公司和互联网服务提供商之间的权力也在发生变化,造成对网络中立制度性支持力度减弱了。

2017年5月,Netflix公司("网络中立"原则坚定支持者之一)首席执行官里德·黑斯廷斯(Reed Hastings)指出,"网络中立"已不是Netflix公司重点考虑的事项。他声称:"10年前,网络中立对Netflix公司非常重要。如今,尽管网络中立对于社会、创新、企业家仍然非常重要,却已不再是我们关注的焦点。"

二、请将下列文本译为英语。

【参考译文】

Apple Inc. is a high-tech company in the United States. It was established by Steve Jobs, Steve Wozniak and Ron Wayne on Apr. 1, 1976, as Apple Computer Inc., and renamed Apple Inc. on Jan. 9, 2007, with headquarters located in Cupertino, California.

Apple Inc. shares went on the public market on Dec. 12, 1980 and setan market value record of 623.5 billion dollars in 2012. By June 2014, Apple Inc. had become the company with the world's largest market value for three consecutive years. Apple Inc. ranked 9th among the world's top 500 in 2016. On Sep. 30, 2013, Apple Inc. overtook Coca Cola and became the world's most valuable brand according to the Best Global Brands report made by Omnicom Group. In 2014, Apple brand overtook Google and became the world's most valuable brand.

On Jul. 20, 2016, *Fortune* released the latest world's top 100 list on which Apple Inc. ranked 9th.

At 1:00 a.m. on Sep. 8, 2016 (Beijing time), 2016 Apple's Fall Launch Event was held in the Bill Graham Civic Auditorium in San Francisco. In Oct., Apple Inc. occupied first place among the global top 100 most valuable brands in 2016.

At 8:00 a.m. on Jan. 6, 2017, the "Red Friday" sales promotion event was formally launched on Apple's official website, and in a flash a large number of users flooded onto the official website for panic buying. On that occasion, all the sales promotion earphones were snapped up within a mere two minutes, and the following notice was displayed (or "appeared") on the official website: "due to a great demand, the Beats Solo3 Wireless headphone is not included in the purchase". Meanwhile, Apple's official website crashed, and the page was unavailable for a time.

In Feb. 2017, Brand Finance released the 2017 global top 500 brands list, and Apple Inc. ranked second. On Jun. 7, 2017, *Fortune* released the 2017 American top 500 list, and Apple Inc. ranked 3rd. On Jul. 20, Apple Inc. ranked 9th on the 2017 global top 500 list.

第 10 章

一、请将下列文本译为汉语。

【参考译文】

电视瘾——一个越来越严重的问题

在我们这一代人当中,大部分人很难想象没有电视的生活是什么样子,因为客厅里闪烁的电视屏幕曾伴随我们长大成人。对我们来说,真的很难看出电视对我们的生活有什么不良影响。很多人认为电视阻碍了亲人和朋友之间的交流。但是,从另一方面来看,电视加快了信息传播的速度,可以将上几代人没有听过的故事和没有见过的图片传递到千家万户。人们往往喜欢美化过去,总是想象着以往的日子里亲人和朋友们围坐在温暖的壁炉前玩游戏、讲故事那其乐融融的场景,但我认为亲人和朋友之间缺乏沟通跟电视没有太大关系。

毫无疑问,与其他交流方式(如互联网、电话、电影、手机,当然了,还有面对面沟通)相比,电视是过去几十年间最强大的交流方式之一。由于其实用性强且内容丰富(有图像和声音),我们真的很难不看电视。在我们生活中,看电视就像穿衣、吃饭和回家睡觉一样必不可少。

首先,广播电视产业是为了公共目的而设立的。如今,提到电视的影响,有人说电视为我们提供了一段放松的时间以及很多有用的信息,还有人说电视对我们也有负面影响,因为电视阻碍了人与人之间的交流,但我觉得电视为我们提供了很多可以谈论的话题以及与亲人、朋友相聚的机会。最重要的是,电视有助于我们与别人交流。几乎每天晚上,我们都跟家人一起看电视。比如,我跟我爸爸喜欢一起看新闻或知识内涵丰富的纪录片节目。看这些节目时,我们会讨论时事,比如总统选举或者经济问题。同样,上大学时我跟朋友最常谈论的也是电视上的热点话题(如新闻或流行节目)。如果前一天晚上有人没有看电视,他们就很难参与到讨论中来,因为他们根本就不知道要说什么。在这一方面,电视为我们在交谈时提供了各种各样与我们生活息息相关的话题,有利于形成我们与家人或朋友之间的亲密关系。

其次,电视可促进人与人的交流,因为电视为我们提供了一个与亲人和朋友相聚的机会。如今,大部分人都比较忙碌,没有太多时间与家人和朋友相处。但是,通过电视节目,我们就可以弥补这个遗憾。例如,2002 年韩国举办世界杯的时候,很多人跟自己的家人和朋友一起度过了一段欢乐的时光。为了看足球赛,我们跟疏远了很久的朋友或家人相聚在电视机前。因而,这类电视节目有利于我们建立更紧密的人际关系。因此,我觉得电视在促进与亲人或朋友交流方面扮演着重要角色。

"儿童更容易积极参与他们感兴趣的事情。尤其是电视,因为电视上各种各样的声音和图像,对于儿童来说极具吸引力。不管电视内容太简单还是太难以理解,他们都渴望看电视。电视上有一些挑战性节目有助于孩子提高词汇量。就像人体的肌肉一样,若经常用脑,大脑则会变得越来越聪明;若长时间不用大脑思考,大脑就会退化。人们总是觉得看电视会对儿童产生消极影响,但是有时我们却忽视了前面所述的这样一个重要事实。"(《科学日报》,2001年)

有些电视节目实际上能够促进儿童的智力发展。电视上有众多类型的节目和内容。有人曾做过的一个研究结果表明,看电视对儿童的影响取决于节目内容。

电视对儿童总是存在不良影响。喜欢看电视的儿童会愈加把生活看作一场娱乐盛宴,他们喜欢在电视节目上扮演角色。现在的电视节目中有一些攻击性内容,对儿童影响很大。另外,电视上不可避免地会出现令人讨厌的暴力画面。我想再一次强调,看电视对儿童的影响取决于节目内容和节目类型。家长应该告诉孩子什么才是他们可以观看的电视节目。

(Carter Bill,1996年)

二、请将下列文本译为英语。

【参考译文】

　　Urban green space, an important medium for urban biodiversity and a key indicator reflecting the quality of urban life, plays a vital role in ensuring the sustainable development of the urban ecological environment and maintaining the physical and mental health of residents. As an important component of the green space system, soil is not only the basic site of growth medium for vegetation, but also a necessary link in the matter to energy transformation cycle of the urban ecosystem.

　　In recent years, some scholars have selected indicators that characterize soil quality, and used high-level comprehensive and multi-index quantitative means to establish a comprehensive index and evaluation system for soil quality. They have also studied the laws of soil evolution in space and time; however, these studies focus more on agriculturally cultivated soil, and less on the evaluation of the physical properties of urban green space soil.

　　Unlike in the case of agriculturally cultivated soil and forest grassland soil, new urban green land soil is more frequently and more severely affected by human activities. These activities ultimately have a cumulative effect on the spatial and temporal variability of the soil's physical composition, and cause an indiscernible degradation of the soil's quality. The physical degradation of the soil restricts both the space for plant roots to spread and their growth rate, and determines plant resistance in an adverse environment, characterized by the dysfunction of the urban green space ecosystem, the effects of the monotony and similarity of the landscape, the degradation of biodiversity, etc.

Soil quality reflects the capacity for interaction between the internal functions of the soil and the external environment. The ability of soil to coordinate nutrients with the environment depends primarily on soil colloids and soil agglomeration, i. e. the physical state of the soil. If the soil quality of urban green space is evaluated based only on soil fertility, then it will not possible to highlight the particular nature of the evaluation objectives. Consequently, the evaluation results will be skewed. Karlen et al. believe that since soil quality evaluations differ as to space, time and particular environment, the establishment of different soil quality evaluation index systems is needed to identity problem areas. Different indicator systems should be used for lands in different regions and those of different landscape types so as to resolve the problem of the appropriateness of the index system, evaluation methods and research scales used.

Taking the Xi'an Botanical Garden, a new park, as a typical case, the author both examines the physical properties of the park's soil and proposes, using fuzzy mathematics, to evaluate the physical quality of the soil in the new green space from the perspective of the spatial variability of the soil's physical properties. S/he also discusses the differences between the soil layers in the new green space and analyzes the status quo of each factor and the interrelationship among them in orderto provide a necessary reference for the systematic evaluation of the physical characteristics of new urban green space in Xi'an City, and then offer basic data for the sustainable development of this city's urban ecological environment and sponge city construction.

第11章

一、请将下列文本译为汉语。
摘要1：
【参考译文】

波兰语发音中的送气现象：这是正在发生的发音变化吗？

×××

摘要

我们通常认为，在重读元音之前，/p/t/k/的送气式发音是典型的英语传统发音，而并非波兰语的传统发音。在波兰语-英语双语话语中，英语和波兰语发声起始时长受语音通用效应、文体和态度等因素的影响，送气是其标志之一。最近，波兰语中的"送气"用法似乎并不表示话语带有英语口音，而表示强调意味。本研究采用话语重点关注范式，从风格调节角度讨论波兰语中长发音起始时间值的使用情况，并讨论当前观察到的变化和基于语音发展总趋势而做出的预测之间的关系。在普遍的语境延伸情况下，我们观察到存在噪声起始时间值延长的趋势，对话语的关注加剧了这一趋势的发展。这表明波兰语中的送气发音可能具有显著的功能。研究中出现的有趣的问题是这种变化是否受语言使用经验的影响，以及此类变化是否会从双语/高级英语学习者扩散至单语者。

摘要 2：

【参考译文】

语篇类型在非文学翻译教学中的应用

×××

摘要

　　语篇类型与翻译（尤其是具有明显文本差异的中英非文学作品翻译）实践相关，但在我国高校翻译教学中，对文本类型相关问题普遍重视不够。以文学作品翻译为中心，中国翻译培训长期以来一直注重词汇和句法层面的技巧，尚未采用基于文本的训练方法，这与给学生布置非文学翻译任务这一明显趋势相吻合。学生在词汇和语法层面上表现出极强的变通能力，在语篇层面上做出的翻译决择却显得牵强附会。尽管学生直观地认识到这两种语言之间存在文本差异，但由于没有经过良好的训练，不会有效处理这些差异，导致翻译文本无法实现其交际功能。本文试图探究这个问题，强调在翻译教学中采用语篇类型的必要性。此外，为了系统地教授语篇类型知识，本文提出了一个新的教学框架。

二、请将下列文本译为英语。

摘要 1：

【参考译文】

Empirical Research of College Teachers' Computer Self-efficacy and Computer Attitude

×××

Abstract

　　Previous studies indicate that college teachers' computer self-efficacy and computer attitude are significant variables which affect the integration of information technologies into the curriculum; thus, it is necessary to discuss the relationship between the computer self-efficacy and computer attitude of teachers. This study selects teachers from 4 colleges in Shandong as respondents, and analyzes the correlation of various dimensions of college teachers' computer self-efficacy and computer attitude. The results show that the correlation of various dimensions of college teachers' computer self-efficacy and computer attitude are statistically significant, and that the perceptive control and behavioral factors of two dimensions of computer attitude are the best predictors of teacher computer self-efficacy.

摘要 2：

【参考译文】

Enlightenment of Evaluation Theory on Text Translation

×××

Abstract

　　This paper examines the contributions of evaluation theory to translation studies. As a development of systemic functional linguistics in keynote theory, Evaluation Theory focuses on research done on attitudinal resources, source of attitude and textual posture in texts. Our findings indicate that translation theory has always attached importance to research on the attitudinal meaning in texts, but has lacked operable tools. In our opinion, Evaluation Theory can be used as a tool for attitudinal analysis in translation. Moreover, Evaluation Theory has implications for the study of textual dialogism and for translation studies. The introduction of Evaluation Theory into translation studies will promote the development of translation theory and translation teaching.

第 12 章

一、请将下列文本译为汉语。

【参考译文】

属性

由于计算机网络依赖相关学科（电气工程、电子工程、电信、计算机科学、信息科学或计算机工程等）的理论和实践应用，所以计算机网络可视为这些学科的一个分支。

计算机网络促进了人际交流，用户可通过不同方式（电子邮件、即时消息、在线聊天、网络电话、视频电话和视频会议等）进行高效便捷交流。用户可以通过网络共享网络和计算资源，访问和使用网络中设备提供的资源（例如，使用共享网络打印机打印文档或共享存储设备等）。网络允许用户共享文件、数据和其他类型的信息，用户也可通过授权访问网络中其他计算机储存的信息。分布式计算通过使用网络中的计算资源完成任务。

黑客可能通过网络连接设备部署计算机病毒或蠕虫，或通过拒绝服务攻击阻止设备访问网络。

网络数据包

不支持数据包传输的计算机通信链接（如传统的点对点式电信链接）仅通过位流传输数据。但是，计算机网络中的大部分信息均以数据包形式传输。网络数据包是一种格式化的数据单位（一串位或字节，通常包含数十个或数千字节），通过数据包交换网络传输。数据包通过网络发送至目的地。传输至目的地后，数据包重组为原始消息。

数据包由两种数据构成：控制信息和用户数据（有效载荷）。控制信息提供网络所需数据以传输用户数据。比如，这些网络所需数据有发送源网络地址和目的地网络地址、错误检测代码和序列信息。控制信息通常位于数据包的包头和包尾，而有效载荷数据位于包头和包尾之间。

与电路交换网络相比，数据包可使用户更好地共享传输介质的带宽。当某一用户未发送数据包时，其他用户发送的数据包可占用链接网络。因此，在链接未被过度使用的情况下，用户可分担成本，干扰也相对较小。通常没有立即可用的数据包传输线路，需排序等待链接网络再次空闲之后才能传输。

二、请将下列文本译为英语。

【参考译文】

Computer System

Introduction

A computer system is a machine system that receives and stores information at the request of an operator, automatically performs data processing and calculating, and outputs resultant information. A computer, as the extension and expansion of brain power, is one of the great achievements of modern science.

The computer system consists of a hardware system (subsystem) and a software system (subsystem). The former, which is the entity on which the system works, is an organic combination of various physical components based on the principles of electricity, magnetism, light and machinery; the latter is a variety of programs and files used to direct the whole system to work according to specified requirements.

Since the time the first electronic computer came out in 1946, computer technology has made remarkable progress in the areas of components and devices, hardware system structure, software systems, applications and so on. Modern computer systems, ranging from microcomputers and personal computers to giant computers and their networks — with their various forms and characteristics — have been widely used in scientific computing, transaction processing and process control. With each passing day, modern computer systems are penetrating ever more deeply into all areas of society and are thus having a profound impact on society's progress.

Electronic computers fall into two categories: digital and analog. The computers that we usually refer to are digital computers. The data processed by such computers is expressed in discrete digital quantities while that processed by analog computers is represented by a continuous analog quantity. Compared with digital computers, analog computers are faster and their interface with physical equipment simpler. On the other hand, they are less precise, less stable, and not as reliable. What's more, such computers are expensive and difficult to use. As a result, analog computers, which are still used occasionally when a rapid, though low accuracy, response is required, are being phased out. Hybrid computers, which combine the advantages of the two, have a certain degree of life force.

第13章

一、请将下列文本译为汉语。
【参考译文】

约克市旅游指南

作为北约克郡的一颗明珠,约克市正在掀起一场超越罗马和维京文化的新浪潮,亦即通过创新型音乐和新颖的餐饮店,打造一个充满活力和创造力的新城市。

提到约克市,你会想到什么?大教堂、蒸汽机、罗马人和维京人,还是一个拒绝21世纪到来的传统城市?透过华丽花哨的外观,在狭窄的中世纪街道之外,约克市有着完全不同的一面。

网络杂志《约克艺术》的编辑丹妮尔·巴尔赫声称:"毫无疑问,约克市正变得更有活力。

近年来,当地许多居民开始经营独立业务:开设小剧场、电影公司、艺术工作室、音乐发烧吧等。人们几乎都在宣泄着叛逆艺术。我们经常可以听到他们说'如果还有其他人尚未尝试的东西,我们自己来试试'。"

当地 DJ 和音乐发烧友绍尔·皮特森(Tor Petersen)建议:放一张"耳虫唱片公司(Earworm Records)"的唱片,几分钟后您将很快融入当地音乐氛围——从"乌洛波洛斯(Ouroboros)""年轻暴徒(Young Thugs)""靛蓝_303(Indigo_303)""阿尼麦克斯(Animaux)""请你(Please Please You)"等音乐中感受当地文化(很多来约克的游客可能都没有意识到这一点)。电子音乐节"坏坏的查普尔(Bad Chapel)"共同发起人彼得森声称:"尽管这是一个小城市,但这里有一群知识渊博、乐于合作的人。在缺乏场地时我们互相帮助。我们为梦想奋斗,感恩拥有的一切。"

从新兴的草根艺术"巴尔赫艺术项目"(Arts Barge Project)到近期即将实施的"点燃约克市"(Spark York)项目(通过提供美味食品和艺术服务的游船,向游客展示不同的社区文化),约克市似乎意识到,通过历史文化类旅游景区吸引游客到来数年以后的今天,当地居民和游客则对这个城市有着更多的期待。食品行业同样如此。尽管约克市仍然主要通过连锁店和"游客陷阱"场馆向游客提供美食,但拥有个性且心怀梦想的当地主要独立经营食品企业(如 Skosh 和 Le CochonAveugle)已使这座城市名扬四方。约克市正在成为美食爱好者的梦想之地。例如,备受当地欢迎的大厨库克·科克里尔(@lukecockerill)当前正在做一个新约克市项目。

但是,约克市的发展仍存在种种障碍。无党派人士〔如福斯盖特星期日(Fossgate's Sunday)街头派对发起人等〕在面对外界无情的竞争压力时,往往会互相帮助以渡过难关。

"餐叉上的约克"(yorkonafork.com)网站编辑本·索普(Ben Thorpe)声称:"我在2001年就已搬到约克市,但在七年之后才发现第一家像样的餐厅。现在,约克市内的餐厅越来越多,但是商业费率过高,连锁店之间存在不公平竞争,我采访过的无党派人士对此均表示不满。"

二、请将下列文本译为英语。
【参考译文】
"The First Insect Bread in the World Launched" in Finland, Containing 70 Crickets
Nov. 25, 2017 08:49:20   From: China News (Beijing)

Foreign media reported that Fazer, a baked goods company in Finland, has put a kind of bread which it claims is the "World's first bread made from insects" on the market. One loaf of bread costs 3.99 Euros (equal to 31 RMB), double thepric of regular bread.

It's reported that this kind of bread is made from flour mixed with minced dried crickets, and that each loaf contains 70 crickets.

Fazer said that this bread contains more protein than regular bread. What's more, the insects are rich in fatty acids, calcium, iron and vitamin B12, which is good for the human body. However, they did not explain the source of the ingredients.

Sarah, a student from Helsinki, said after she had a taste: "I don't feel there is any difference. This bread tastes just like regular bread."

Scherbekov, the R&D supervisor of Fazer, observed: "Cricket bread is a great source of protein, and it might allow (help) consumers to become increasingly familiar with foods made from insects." It is predicted that gradually the market for this bread will be extended to 47 branch stores.

第 14 章

一、请将下列文本译为汉语。

【参考译文】

非法伐木工因非法占有森林产品遭到逮捕并处罚金

贝尔莫潘:2017 年 9 月 28 日

昨日,橘园区羊草村 47 岁居民帕布洛·维拉戴瑞兹(Pablo Valladarez)因非法持有桃花心木被判处罚金 22 838 美元。2017 年 9 月 24 日,警察和林业部门官员在橘园区老北方高速公路上拦截了羊草村和圣玛莎村居民维拉戴瑞兹(Valladarez)、卡塔里诺科波(CatalinoCopo)、尼尔森·默萨(Nelson Mosa)和奥雷里奥·麦(Aurelio Mai)等四人,并拖走他们非法采伐的桃花心木。

在这次行动中,林业部门共没收 34 根未经批准采伐的桃花心木原木(相当于 1 268.8 板尺),根据 2000 年颁布的《森林法》和 2017 年颁布的《森林法》修正案,起诉上述四人犯有"非法持有林木产品"罪。9 月 27 日,橘园区地方法院法官艾尔伯特·霍尔(Albert Hoare)对该案件开庭审理,维拉戴瑞兹等人对所控告的罪行供认不讳。根据判决,如果维拉戴瑞兹未能在 2018 年 1 月 31 日之前缴清罚款,他可能会面临五年监禁。

在该事件发生之前,林业部门曾在橘园区卡梅丽塔村居民胡安·雷耶斯(Juan Reyes)处截获多种盗伐的木材(如尼格利陀、圣玛利亚和桃花心木等)。2017 年 9 月 19 日,雷耶斯因非法持有森林产品最终被处以罚款。

橘园区林业部门负责人索尔·克鲁兹(Saul Cruz)声称:"林业部门正在严厉打击非法采伐这一普遍存在的违法行为"。克鲁兹指出,尽管木材采伐利润丰厚,但盗伐木材破坏了当地的发展机会。克鲁兹继续补充道:"非法采伐这一林业犯罪行为不仅会对当地伐木业的可持续发展和我们生存的环境造成威胁,还会影响靠森林为生的居民生计,甚至我国经济。"

鉴于违法行为的严重性,2017 年颁布的《森林法》修正案明文规定了更严厉的处罚措施,但林业部门认识到只靠法律约束远远不够。因此,林业部门将继续通过加强和其他执法部门之间的协作、建立社区保护意识并增强监察和执法工作等措施加大管治力度。

林业部门提醒民众,6 月 15 日至 10 月 15 日为禁伐期,在此期间严禁采伐。该措施不仅可以确保更加安全的伐木作业,保护森林土壤,而且还能减少雨季对公共道路的破坏。民众可通过拨打 822-2079(电话号码)或发送邮件至 info@forest.gov.bz(邮箱地址)联系林业部门,举报林业犯罪行为。

二、请将下列文本译为英语。

【参考译文】

Xi Jinping and Mrs. Peng Liyuan Hold Banquet to Welcome President Trump and His Wife

Li Keqiang, Zhang Dejiang, Yu Zhengsheng, Zhang Gaoli, Li Zhanshu, Wang Yang, Wang Huning, Zhao Leji, Han Zheng, Liu Yunshan and Wang Qishan attended the banquet.

As reported by the Xinhua News Agency on Nov. 9 in Beijing (reporters Tan Jingjing and Xu Ke), President Xi Jinping and Mrs. Peng Liyuan held a banquet in the Great Hall of the People on the evening of the 9th day to welcome U. S. President Trump and his wife Melania. Li Keqiang, Zhang Dejiang, Yu Zhengsheng, Zhang Gaoli, Li Zhanshu, Wang Yang, Wang Huning, Zhao Leji, Han Zheng, Liu Yunshan and Wang Qishan (also) attended the banquet.

During the banquet in the splendid and magnificent Great Hall of the People, where all seats were occupied by eminent guests, both heads of state delivered warm, enthusiastic speeches.

Xi Jinping pointed out that although there is a vast ocean between China and the United States, geographical distance has never been a barrier to these two great nations coming together and that neither side has ever ceased in its efforts to pursue friendship and mutually beneficial cooperation. Forty-five years ago, President Nixon visited China, a visit that reopened the door for exchanges between China and the United States. Since then, thanks to the joint efforts of several generations of leaders and the people of our two countries, historic progress has been made in China - U. S. relations. It has benefited our two peoples and has changed the world. Today, China-U. S. relations have grown into a community with coinciding interests. Now, at this point in time our two countries share far more and broader common interests than before and must shoulder greater and weightier responsibilities in the form of upholding world peace and promoting common development. The strategic significance and global influence of China-U. S. relations continues to grow.

Xi Jinping emphasised that this visit by President Trump to China was an event of historic importance: "Over the past two days, President Trump and I have mapped out a blueprint for the future development of China-U. S. relations. We both agree that China and the United States should remain partners, not rivals. We both agree that when we work together, we can accomplish many great things to the benefit of our two countries and the whole world. As an old Chinese adage goes, "no distance, not even remote mountains and vast oceans can ever prevent people with perseverance from reaching their destination". It is my firm conviction that although China-U. S. relations face limited challenges, the potential for growth is boundless. With perseverance, we can surely write a new chapter in the history of China-U. S. relations. With perseverance, our two great nations will definitely be able to make new contributions to the creation of a better future for mankind."

Trump expressedhis gratitude to president Xi Jinping for the warm hospitality, adding, "The people of the United States have a very deep respect for the heritage and traditions (culture) of China. In this historic moment, I deeply believe that American-Chinese cooperation will benefit the people of both countries and bring peace, security and prosperity to the world."

Before the banquet, classic highlights from the meetings between President Xi Jinping and President Trump in Mar-a-Lago and Hamburg as well as highlights from the president's current visit to China were shown in the golden hall. President Trump then proposed that a video showing his granddaughter, Arabella, singing traditional Chinese songs and reciting classical Chinese poetry along with the Three Character Classic be played. The Great Hall burst repeatedly into warm applause.

In addition to the above mentioned, those present at the banquet included Ding Xuexiang, Wang Chen, Liu He, Liu Yandong, Yang Jiechi, Guo Shengkun, Cai Qi, Han Qide, Dong Jianhua, Wan Gang, Zhou Xiaochuan as well as U. S. Secretary of State Tillerson, other members of the cabinet, and high ranking White House officials. (Finished)

Executive editor: Wang Minhe

第 15 章

一、请将下列文本译为汉语。
【参考译文】

远程医疗进入家庭

医学：通过视频链接，远程医疗可提供远程问诊甚至远程手术，但其未来发展可能与家庭关系更为密切。

特里斯坦达库尼亚群岛几乎是地球上最与外界隔绝的地方。这座火山岛位于南大西洋中部，距离南非1 750英里，距离南美洲2 088英里，是全球最偏远的人类居住点，岛上仅有269位居民。若当地居民患有不常见疾病或遭受严重伤害都是棘手问题。卡雷尔·范德尔莫维(Carel Van der Merwe)是该岛上唯一的医生。他声称，"该岛没有飞机跑道，因此无法将患者快速转移到其他地方去救治。该岛若与外界接触至少需要六至七天的远洋航行，因此所有医疗措施均需要在群岛上完成。"

尽管如此，受益于特里斯坦项目(精心设计的远程医疗实验项目)，该岛上的居民现在可利用世界上一些最先进的远程医疗设施治疗疾病。通过电信和医疗技术手段，该领域(远程医疗)正随着科技进步而快速发展。首先，远程医疗(例如，向专家发送电子文本格式的X光图片)帮助医生和医务人员交换信息。此类应用现在已越来越普遍。全球计算机巨头 IBM 公司是参与特里斯坦项目的其中一家公司,其首席医务官理查德·巴加拉(Richard Bakalar)声称:"(在这里)我们已经建立了一种患者-医生救治模型。"

通过卫星-互联网连接美国的 24 小时紧急医疗中心,范·德尔·莫维(Van der Merwe)医生可向医疗专家发送数字格式的 X 光图片、心电图(ECGs)和肺功能检测文件。如果需要,他可通过视频咨询相关的医疗专家。通过该系统,心脏病学专家甚至可以从地球的另一端测试和重新编程心脏起搏器程序,或对已植入患者体内的除颤器编辑程序。总之,当范·德尔·莫维医生对患者实施外科手术时,患者相当于在匹兹堡大学医疗中心接受手术治疗。当地居民了解专家咨询意见之后,可以安心地在岛上居住。

IBM 医疗卫生专家保罗·格兰迪(Paul Grundy)声称,远程医疗所需的大部分技术现在均可实现,建立相应设施也非常简单。他说,远程医疗最大的困难是安装卫星-互联网连接线路。理论上讲,此类远距离电信医疗技术发展的空间很大。2001 年,纽约市的一名外科医生使用名为"达·芬奇"的机器人外科手术系统,成功为巴黎市的一名患者实施胆囊切除手术。尽管该技术给人留下深刻印象,但并不属于远程医疗的主流发展方向。

家庭是远程医疗技术的重点发展方向。

在远程医疗发展方面,远程技术的进步速度落后于医疗技术。从长远来看,远程医疗不会仅局限于为身体不适的患者提供远程医疗护理,而是通过穿戴式或植入式传感器监控患者,在早期诊断出疾病。远程医疗技术的发展重点将从急性病转移向慢性病,从治疗转移至预防。如今,利用远程设备为患者提供医疗服务只是远程医疗的一种应用方式,但与更广泛的应用领域相比,该应用方式仅属于冰山一角。远程医疗技术最终将逐步进入家庭。

二、请将下列文本译为英语。

【参考译文】

Natural products are metabolites synthesized by natural organisms during their life activities. Drugs derived from such products have been widely used to treat a variety of major diseases such as cardiovascular diseases, malignant tumors, immunological diseases and infectious diseases. Compared with chemically synthesized small molecule drugs, natural products, which have been naturally selected and optimized during long-term evolution, have obvious advantages as to structural novelty, biocompatibility, functional diversity and other features.

Natural products and their derivatives account for a large proportion of the research on and development of new drugs and clinical medication. According to statistics, a considerable number of the listed drugs approved for the market by the US Food and Drug Administration (FDA) from 1939 to 2016 have either contained molecular fragments (more than 50%) from natural products, or been directly derived from such products. Therefore, the use of natural products for drug design is an effective way to generate modern innovative drugs, based on the interaction of active natural products with intracellular targets.

The first key step in the discovery of new drugs based on natural products is to identify the targets of drug molecules at the initial stage of drug research and development. This helps in the further study of their active mechanisms and in discovering possible toxic side effects as early as possible, thereby enabling specific structural changes to be made while reducing the cost of drug research and development.

In an organism, phenotype changes may be brought about at different levels e. g., at the genome, transcriptome, proteome, and metabolome levels, by combining a small molecular compound including natural products with a macromolecular target. Since proteins are the major performers of cellular functions, macromolecular targets are, in most cases, proteins that interact with natural products, i. e. target proteins. Based on the logical relationship between small molecule, target, and phenotype, strategies for the identification of targets for natural products can be divided into two categories as follows: One is to screen target proteins using small molecule natural products as the starting point. This is known as the reverse strategy, and includes methods such as chemical proteomics, chemical genomics, and biophysics; the other is to speculate on (infer) and identify targets of natural products based on phenotype changes caused by natural products or by known signal pathways and interaction networks. This is known as the forward strategy, and mainly includes differential genomics/proteomics analysis, cell morphology analysis, etc. Using representative examples from the past two decades, this paper will introduce some methods and strategies for the identification of such targets.

第 16 章

一、请将下列文本译为汉语。
【参考译文】

润滑系统

往复式发动机使用湿式油底壳或者干式油底壳润滑系统。在湿式油底壳润滑系统中,润滑油位于油底壳中。油底壳是发动机不可分割的一部分。在干式油底壳润滑系统中,润滑油存储在独立的油箱中,通过油泵在发动机里循环。

湿式油底壳润滑系统的主要部件是送油泵,送油泵从油底壳抽出润滑油并引流到发动机中。润滑油流经发动机后,返回油底壳。在一些发动机内,旋转的曲轴把润滑油飞溅到发动机的各个部分,为发动机提供了额外的润滑作用。

干式油底壳润滑系统中,送油泵也提供油压,但是油源位于发动机外部独立的润滑油箱。润滑油流经发动机后,会被回油泵从发动机的不同部位送回到润滑油箱。干式油底壳润滑系统能够供给发动机更大容量的润滑油,更适用于大型的往复式发动机。

润滑油压力表直接指示润滑系统的运行情况,供给发动机的润滑油油压以磅/平方英寸为单位。绿色表示油压在正常工作范围,红色表示最小和最大压力。发动机启动时润滑油压力表上有油压指示。请参考《飞机飞行手册》或者《飞行员操作手册》了解飞机制造商给出的限额。

润滑油温度表测量油温。绿色区间表示正常工作范围,红线表示最大允许油温。与润滑油压力不一样,润滑油温度变化更为缓慢,这一点在发动机冷启动时尤其明显,可能需要几分钟或者更长时间温度表上才会显示增加的油温。

尤其当环境空气温度低或者高时,在飞行过程中应定期检查润滑油油温。过高的油温表读数可能表示油管堵塞、油量变少、润滑油冷却器阻塞或者油温表故障,过低的油温读数可能表示在冷空气中运行时,润滑油的黏稠度不正常。

二、请将下列文本译为英语。

【参考译文】

Oil Source Vehicle Operation Procedures
UnauthorizedOperators Are Forbidden to Operate Oil Source Vehicles

Ⅰ. **Pre-operation checks**

1. Power supply voltage:(380±38) V. Three-phase voltage should be balanced.

2. All control switches of the operating console should be turned to "OFF".

Ⅱ. **Operating steps**

1. Turn on the power, then turn the heat pipe to "ON" to heat the lubricating oil.

2. When the oil temperature rises to or near to the desired value, start the scavenge pump and adjust the lubricating oil flow rate to the prescribed range.

3. Start the drive platform, increase the rotation speed of the device under test (DUT) to working speed, and adjust the flow rate and oil pressure to meet the prescribed requirements. If one of them fails to meet the requirements, the test should be discontinued.

4. When the DUT is working properly, observe when the pressure, temperature and flow rate of the lubricating oilreaches the prescribed value, then turn off the heating switch and open the cold water valve, thereby implementing the automatic adjustment of the oil temperature.

5. After the test, slowly reduce the rotation speed of the DUT to zero, then turn the oil flow rate to zero. Wait one to two minutes to allow the lubricant within the DUT to be sufficiently pumped clean before turning off the power of the scavenge pump.

6. Turn off both the operating power of the oil source vehicle and the main power source.

Ⅲ. **Caution**

If anomalies occur during the testing process, reduce the DUT rotation speed to zero, then shut off the power of the oil source vehicle

Ⅳ. **Maintenance**

1. The operating console panel should be cleaned regularly in order to keep the surface clean.

2. Spot inspection instructions:

(1) Before each experiment, make sure that the oil level in the oil tank reaches to the

middle position on the oil scale. Before heating the oil tank, use an oil probe to confirm that the oil level is about 2cm from the top of fuel tank. If not, refuel the oil tank.

(2) Before each experiment, Check whether the filter is clean; all dirt must be removed.

(3) When the pump is started, check the oil flow rate, pressure and temperature as well as whether there are any abnormal vibrations.

(4) Check to see whether there is leakage in any of the joints.

(5) Check to see that the various devices and meters display accurately.

3. Make sure that the oil flow rate, pressure, and temperature sensors are measured and calibrated in accordance with the relevant regulations.

第 17 章

一、请将下列文本译为汉语。

【参考译文】

英国白皮书发布产业战略 促进生产力发展

英国商务大臣宣布建立新监管部门,以监控白皮书中设定目标(如生物科技投资目标)的实际进展情况。

格雷格·克拉克(Greg Clark)认为,为促进英国本土企业成为新技术浪潮(如机器人和基因组学)的领军者,政府应扮演一定角色。摄影:安东尼·厄普顿/PA(图下英语汉译)

英国政府将在周一发布新产业战略具体细节,旨在提高英国生产力,指明英国需要提高业绩的五个关键领域。

英国商务大臣格雷格·克拉克将宣布成立独立的监管机构,监控英国在推动创新、升级基础设施、提高工人技能水平方面的进展情况,确保通过企业资金水平反映城市实力,进而实现全国的繁荣发展。

克拉克即将颁布的白皮书还将揭示在四个具有增长潜力的行业(建筑、生命科学、汽车和人工智能),英国政府计划与相关企业建立长期战略合作伙伴关系。

克拉克声称,美国生命科学企业默克公司(Merck)(在欧洲,该公司也称为MSD)在英国的新投资创造了950个岗位。这一投资表明,英国政府、私营部门和大学之间建立的合作关系能为英国带来各种福祉。英国政府希望在上述四个行业领域建立这种合作关系。

"生命科学一直是英国发展最快的行业之一,该行业的营业额已超过640亿英镑,为英国提供了233 000个工作岗位。

"默克公司现在的承诺以及我们在更多行业内获得的投资都证明,产业战略中概述的流程可给企业在英国投资提供必要的信心和方向。产业战略将确保英国继续处于创新的前沿,这也表明我们对产业战略拥有坚定的信心。"

白皮书重点关注今年年初发表绿皮书以来经广泛磋商确定的五个核心领域。克拉克认为,上周提出预算案致使英国对产业战略的需求更加迫切。英国预算责任办公室将英国长期生产率增长趋势预测结果减半,下调至1%。白皮书指出,由于脱欧公投,英国更需要抓住机会发展自身。

二、请将下列文本译为英语。

【参考译文】

<h3 style="text-align:center">Business Cooperation Agreement</h3>

Party A：

Legal representative：

Address：

Postal code：

Contact TEL.：

Party B：

Legal representative：

Address：

Postal code：

Contact TEL.：

In accordance with the principle of equality, free will and mutual benefit, Party A and Party B have reached the following agreement through friendly negotiations in order to form a long-term business relationship of cooperation：

Ⅰ. Duration of cooperation agreement

This agreement is valid for five years, from MM/DD/YY to MM/DD/YY.

Ⅱ. Contents of cooperation agreement

Rights and obligations of Party A

1. Party A shallrecommend suitable customers or projects to Party B.

2. Party A shall assist Party B infacilitating contract signing with customers.

3. If Party A has not signed a contract with a customer, Party B's cooperation process shall be subject to Party A's arrangements.

4. If a customer recommended by Party A to Party B signs a contract directly with Party B, Party A will not bear any responsibility.

Rights and obligations of Party B

1. Party B shall recommend suitable customers or projects to Party A.

2. Party B shall assist PartyA in facilitating contract signing with customers.

3. Party B shall provide best services and the most favorable prices to customers recommended by Party A.

4. Party B agrees to effect settlements in accordance with the agreed upon (stipulated) time and methods.

Ⅲ. Cooperation agreement conditions

1. If a customer recommended by Party A to Party B or Party B to PartyA signs a contract, it shall be deemed to be a successful recommendation.

2. After a project has been successfully recommended, the party to whom it has been

recommended shall pay 2% of the actual turnover from this project to the recommending party as a reward.

3. Method of payment: The receiver of the recommendation shall begin making payments from the second month after receiving the customer service fee for the first time, and thereafter payments shall be made once a month before the 5th day. The recommending party shall provide the corresponding official invoice.

Ⅳ. Responsibility for breach of contract

1. If the customer's business reputation or customer relationships are damaged by one party in the process of implementation, the other party may unilaterally cancel the cooperation relationship immediately. Meanwhile, the damaged party will no longer be required to pay related expenses due for the unfinished business, while the other party shall continue to fulfill payment obligations.

2. During profit distribution, if one party raises an objection to the base number and manner of profit distribution, an accounting firm shall be hired to conduct an audit.

Ⅴ. Supplements and changes

For the supplements and changes requested by both parties in the process of implementing this agreement, they can make a supplementary agreement, which has the equal force of law. If there are discrepancies between this agreement and the supplementary one, the latter agreement prevails.

Ⅵ. Termination of this agreement

1. If either party wants to terminate this agreement, the party that makes the proposal shall inform the other party one month in advance.

2. When this agreement expires, both parties shall give priority to the renewal of cooperation with each other.

Ⅶ. Handling disputes

In case of disputes, they shall be settled by both parties through active negotiation. If negotiation fails, the damaged party can file a lawsuit at the People's Court of Guangzhou.

Ⅷ. Effectiveness of this agreement

The agreement shall take effect after being sealed by both parties. Both parties will receiveone of two copies of the agreement, each having equal force legally.

Party A:

Signature of the representative:

Date: MM/DD/YY

Sealed:

Party B:

Signature of the representative:

Date: MM/DD/YY

Seal:

第 18 章

一、请将下列文本译为汉语。

【参考译文】

个人陈述

×××

我是伯明翰大学的一名学生,希望在贵校攻读经济学博士学位,攀登学术高峰。在大学初次接触经济学时,我就认识到经济学是一门兼具艺术性和科学性的学科,很有趣。精确的规则、原理和公理明确界定了该领域的范围,但经济学在解释和分析方面却有着极其广阔的自我表达空间。经济学的这一特性深深吸引着我,我决定在该领域继续深造。同时,身为一名理科硕士生,我意识到所学知识远远不够。因此,我希望在伯明翰大学攻读经济学博士学位。

尽管成长经历和普通孩子一样,但是我的童年比较特殊。与中国绝大多数学生不同,我从约七岁开始在家自学。在我父母得知除非我从高中毕业,否则我将无法参加高考之前,我从未上过学。自学让我具有独自克服各种困难的能力和良好的自控力。

通过前面各阶段的学习,我已在数学、计算机科学、经济学理论和博弈论方面打下了扎实的学习基础。攻读学士学位期间,我学习了基本的经济学理论知识。进入伯明翰大学之后,我曾幸运地聆听了宏观经济学、微观经济学和计量经济学领域优秀教授主讲的课程。通过上述课程的学习,我不仅了解到经济学概念和理论知识,而且拓宽了视野,我对经济学领域有了全新的认知。

本科期间,我曾学习了一门名为"运筹学"的课程,我开始对博弈论非常感兴趣。我发现博弈论如此具有魅力,以至于当我在选课清单中看到这门课程时,毫不犹豫地选择它。尽管学习中需面对诸多难题,但我非常喜欢思考和解决博弈论问题。

博弈论是我真正希望探索的领域。我对与中国传统军事专著-《孙子兵法》有关的博弈论建模非常着迷。我认为博弈论模型对分析战争的特点非常有效,有助于我们更好地理解这本书的内容。

经过学术研究中的重重磨练,我已具备独立研究能力。为了更具洞察力,现在我需要在更广泛的理论框架下不断进步,这是我希望在您的研究项目中继续深造的原因。我衷心期待在您的指导下完成学业,也希望在相关课题中做出更大贡献。

二、请将下列文本译为英语。

【参考译文】

Personal Statement

Apr. 22, 2015

To whom it may concern:

Greetings! My name is Liu Jinshun. I was born in Xiacha Village, Liuliang Township, Zhuanglang County in Oct. 1996 to a poor family of farmers. Though I come from a poor family, I am hardworking and eager to learn. I am determined to leave this backward village and enter the civilized, developed urban world.

I was admitted to Zijing Middle School, the second of key senior middle school in Zhuanglang County, with excellent exam scores in Sep. 2012. I made continuous efforts and strove to make progress at school, where I was ranked top among the outstanding students in my studies; I also served as monitor in the practical activity class. Moreover, during my three years in high school, I won the title outstanding student in my school and the honor of outstanding class cadre; I also won a certificate of honor ... issued by the Pingliang Education Bureau. In addition, I won the 4th Campus Civilization Star of Zhuanglang County certificate of honor issued by the Zhuanglang County Party Committee Publicity Department. As a result of these experiences, I have become accustomed to being outstanding, the feeling of which and helped me develop my collective consciousness and become aware of my responsibility to serve the collective. It has also developed my organizing ability of campus activities.

To be honest, I do not have a particular personal specialty, but I clearly like reading and delving into scientific questions. I think it is enjoyable to read, and reading inspires me to make progress and enrich my spiritual world; I find it exciting to ponder over and solve scientific problems. In my future career, therefore, I hope to engage in scientific research in a key industry concerned with the future and destiny of the nation. This is not only out of interest in such research and the enjoyment I get from engaging in it, but also because I have a national consciousness and want to devote my life to making contributions to our country's development.

The major offered by your esteemed school is in accordance with my wishes. Consequently, I eagerly hope to become a student of your school. I will strive to become an outstanding student as well as to contribute to strengthening our military and to national defense security!

Thank you very much for taking the time to review my personal statement.

With best wishes
Self-recommendation: Liu Jinshun

附　录

附录1　翻译补充练习

一、英译汉

1.

Human Cloning

By Kevin Bonsor

Scientists in South Korea claimed to have created human embryos via cloning. This could radically change the medical landscape-therapeutic cloning could be used to combat diseases like Parkinson's and Alzheimer's.

Nothing really prepared the world for the 1997 announcement that a group of Scottish scientists had created a cloned sheep named Dolly. Many folks believe that within the nextdecade, we will hear a more shocking announcement of the first cloned human. Scientists in South Korea have already created human embryonic stem cells through cloning.

Until now, the idea of human cloning has only been possible through movie magic, but the natural progression of science is making human cloning a true possibility. We've cloned sheep, mice and cows. So why not humans? Some countries have set up laws banning cloning, but it is still legal in many countries.

Me, Myself and My Clone

In January 2001, a small consortium of scientists led by Panayiotis Zavos, a former University of Kentucky professor, and Italian researcher Severino Antinori said that they planned to clone a human in the next two years. At about the same time, the New York Post reported a story about an American couple who planned to pay $500,000 to Las Vegas-based Clonaid for a clone of their deceased infant daughter.

These scientists may be chasing glory in the name of science. Whatever their motivation, it's likely that we will see the first cloned human baby appear on the evening news in the next decade. Scientists have shown that current cloning techniques work on animals, but only rarely do they succeed in creating a cloned embryo that makes it through birth.

If human cloning proceeds, one method scientists can use is somatic cell nuclear transfer, which is the same procedure that was used to create Dolly, the sheep. Somatic cell nuclear transfer begins when doctors take the egg from a donor and remove the nucleus of the egg, creating an enucleated egg. A cell, which contains DNA, is then taken from the person who is being cloned. The enucleated egg is then fused together with the cloning subject's cell using electricity. This creates an embryo, which is implanted into a surrogate mother through in vitro fertilization. If the procedure is successful, then the surrogate mother will give birth to a baby that is a clone of the cloning subject at the end of a normal gestation period. Of course, the success rate is only about one or two out of 100 embryos. It took 277 attempts to create Dolly. Take a look at the graphic below to see how the somatic cell nuclear transfer cloning process works.

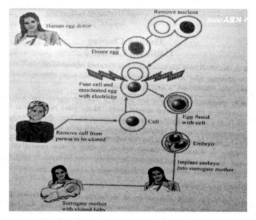

Some scientists seem to think that human cloning is inevitable, but why would we want to clone people? There are many reasons that would make people turn to cloning. Let's explore a few of these reasons.

Who Will Clone?

Not all cloning would involve creating an entirely new human being. Cloning is seen as a possible way to aid some people who have severe medical problems. One potential use of cloning technology would involve creating a human repair kit. In other words, scientists could clone our cells and fix mutated genes that cause diseases. In January 2001, the British government passed rules to allow cloning of human embryos to combat diseases such as Parkinson's and Alzheimer's.

While it may take time for cloning to be fully accepted, therapeutic cloning will likely bethe first step in that direction. Therapeutic cloning is the process by which a person's DNA is used to grow an embryonic clone. However, instead of inserting this embryo into a surrogate mother, its cells are used to grow stem cells. These stem cells can be used as a human repair kit. They can grow replacement organs, such as hearts, livers and skin. They can also be used to grow neurons to cure those who suffer from Alzheimer's, Parkinson's or Rett Syndrome.

Here's how therapeutic cloning works:
- DNA is extracted from a sick person.
- The DNA is then inserted into an enucleated donor egg.
- The egg then divides like a typical fertilized egg and forms an embryo.
- Stem cells are removed from the embryo.
- Any kind of tissue or organ can be grown from these stem cells to treat the sick.

Others see cloning as a way to aid couples with infertility problems, who want a child with at least one of the parent's biological attributes. Zavos and Antinori say that helping these couples is the goal of their research. Zavos said that there are hundreds of couples already lined to pay approximately $50,000 for the service. The group said that the procedure would involve injecting cells from an infertile male into an egg, which would be inserted into the female's uterus. Their child would look the same as the father.

Another use for human cloning could be to bring deceased relatives back to life. Imagineusing a piece of your great-grandmother's DNA to create a clone of her. In a sense, you could be the parent of your great-grandmother. This opens the door to many ethical problems, but it's a door that could soon be opened. One American couple is paying $500,000 to Clonaid to clone their deceased daughter using preserved skin cells.

To Clone or Not to Clone

Critics of cloning repeat the question often associated with controversial science: "Just because we can, does it mean we should?" The closer we come to being able to clone a human, the hotter the debate over it grows. For all the good things cloning may accomplish, opponents say that it will do just as much harm. Another question is how to regulate cloning procedures.

There is no federal law banning cloning in the United States, but several states have passed their own laws to ban the practice. The U.S. Food and Drug Administration (FDA) has also said that anyone in the United States attempting human cloning must first get its permission. In Japan, human cloning is a crime that is punishable by up to 10 years in prison. England has allowed cloning human embryos but is working to pass legislation to stop total human cloning.

While laws are one deterrent to pursuing human cloning at this time, some scientists believe the technology is not ready to be tested on humans. Ian, one of co-creators of Dolly, has even said that human cloning projects would be criminally irresponsible. Cloning technology is still in its early stages, and nearly 98 percent of cloning efforts end in failure. The embryos are either not suitable for implanting into the uterus or they die sometime during gestation or shortly after birth.

Those clones that do survive suffer from genetic abnormalities. Some clones have been born with defective hearts, lung problems, diabetes, blood vessel problems and malfunctioning immune systems. One of the more famous cases was a cloned sheep that was born but suffered from chronic hyperventilation caused by malformed arteries leading to the lungs.

Opponents of cloning point out that while we can euthanize the defective clones of other animals, it's much more morally problematic if this happens during the human cloning process. Advocates of cloning respond that it is now easier to pick out defective embryos before they are implanted into the mother. The debate over human cloning is just beginning, but as science advances, it could be the biggest ethical dilemma of the 21st century.

(韩孟奇. 科技英语阅读[M]. 上海：上海交通大学出版社，2002.)

2.

Speech Acts in Corpus Pragmatics:
A Quantitative Contrastive Study of Directives in Spontaneous and Elicited Discourse
Ilka Flöck and Ronald Geluykens

Abstract

This study compares directives in three different language corpora collected under different conditions: ① spontaneous spoken data (taken from the British component of the International Corpus of English); ② spontaneous written data (viz. business letters), and ③ elicited written data (collected through Discourse Completion Tasks). It is shown that there are significant differences between spontaneous and elicited data sets as well as between spoken and written natural data. These differences occur both in the so-called directive head act as well as in the modification strategies accompanying the head act (downgrading and upgrading), resulting in various levels of directness in the realization of directives in all three data sets. These results show the importance of quantitative comparative research not just across data collection methods, but also across discourse genres, based on corpora of authentic speech.

(FLÖCK I, GELUYKENS R. Speech Acts in Corpus Pragmatics: A Quantitative Contrastive Study of Directives in Spontaneous and Elicited Discourse[C]//TRILLO J R. Yearbook of Corpus Linguistics and Pragmatics 2015: Current Approaches to Discourse and Translation Studies. Switzerland: Springer International Publishing, 2015: 7 - 37.)

3.

Forms of Human Information and Its Communication

Just as there exist different forms of energy — mechanical, chemical, electrical, heat, sound, light, nuclear, etc. — so do there exist different forms of information. Human information represents only one form of information. We will discuss non-human forms later. However, human information itself, may be stored and communicated in a wide variety of ways and represent many different forms.

The systems for storing and processing information contained within the human brain are so complex and so mysterious as to constitute the last great frontier of the biological sciences. When compared to a computer, the human brain exhibits substantially greater complexity in at least three areas (as reviewed by Stonier 1984). First, the circuitry is incredibly more complex: Not only does the brain contain of the order of 10 cells, but a single brain cell may be connected to thousands of other cells. Each neuron, in turn, may prove to be equivalent to atransputer rather than a transistor. Second, the transmission system is different. In a computer one is dealing with electrons moving along a conductor. Nerve impulses, in contrast, involve the progressive depolarization of membranes. This device allows the transmission system to be regulated much more delicately and relates to the third major difference: Information handling in the present generation of computers is digital. In the human nervous system there exist dozens of neurotransmitters and other related substances which can enhance or inhibit nerve impulses-the whole system is a finely tuned and integrated network of analog devices.

The nature of the information inside people's heads must be different from that contained inside a computer. In addition, the form information takes as it is being communicated between two people, two computers, or between people and computers, must differ again. Just as there exist many forms of information, so there exist many means by which information may be transmitted, or transduced (converted from one form to another).

Consider the information on this page. It is being transmitted to the reader's eye by light. The light striking the retina is converted to nerve impulses which are propagated by the sequential depolarization of membranes. At the synapses between the nerve cells of the brain, the information is converted to pulses of chemical neurotransmitters which, in turn, trigger further neurological activities, branching in many directions. Ultimately, these events lead to a host of brain activities: short-term memory storage, comparison with existing information at many levels (from comparing pictures of the printed letters and words, and their meanings, to comparing the concepts of this article with the reader's view of the world), long-term memory storage and the myriad of other, still mysterious, thought processes associated with assimilating and analyzing new information. At some point in the future, the reader may convert the patterns of neural information stored in the brain into sound waves via nerve impulses to the vocal chords. Sound waves, represent a mechanical coding of information. The sound waves impinge upon the ears of the listener where the information is nowconverted from pulses of mechanical energy into nerve impulses by the motion of microscopic hair-like organelles in the inner ear. These nerve impulses enter the brain of the listener where the information will undergo a processing similar to that originally taking place in the brain of the reader.

Alternatively, the reader might speak into a telephone where the information is transduced from patterns of compressed air molecules traveling at the speed of sound into electronic pulses traveling down a copper wire closer to the speed of light. These electronic pulses traveling down awire, might, in turn, be converted into pulses of light traveling down an optical fiber. Or the reader might speak into a microphone either for broadcast, in which case the information is transduced into patterns of electromagnetic waves traversing the airwaves, or into a tape recorder where the electronic pulses are converted into magnetic pulses, then "frozen" into the tape by having atoms, responding to magnetism, arrange themselves physically into patterns of information within the tape.

The above represents one set of cycles which illustrate the communication of human information. Note what it entailed: the information was propagated as:

1. Patterns of light (from the book to the eye).
2. Pulses of membrane depolarization (from the eye to the brain).
3. Pulses of chemical substance (between individual nerves).
4. Pulses of compressed air molecules, i. e., sound waves (emitted by the larynx of the speaker).
5. Pulses of mechanical distortion in liquid or solid (inside the ear or the telephone mouthpiece).
6. Pulses of electrons in a telephone wire.
7. Pulses of light in optical fibers.
8. Pulses of radio waves.
9. Pulses of magnetism (inside the earpiece of a telephone or the speaker of a radio).

The information was stored as printing in a book, in the human brain, and on magnetic tape.

The first involved patterns of dye molecules; the second, probably patterns of neural connections; and the last, patterns of magnetized regions. It was converted from one form into another in the retina of the eye, at the synapsis between nerve cells, in the larynx, in the inner ear, in the telephone, the radio transmitter, the radio receiver, and the tape recorder.

The reader might also have decided to store the information on a computer, or in a file by photoduplicating it, or by typing it out; thereby committing the information once more to patterns of dye molecules superimposed on paper molecules. From cave wall paintings and carvings on wood or stone, to bubble memories and satellite communication, human information is capable of being stored, transmitted or transduced in a very large number of ways-and the number continues to increase.

It is important to note that the means of propagating information, as exemplified by the above list of nine, usually involved pulses of waves (light, sound, radio waves), pulses of electrons, or pulses affecting matter or its organization. The fact that information may be divided up into small, discrete packets is utilized by communications engineers for packet switching to allow several users to use the same facility simultaneously. The idea that information is an independent entity, comes naturally to the communications engineer since the pioneering work of Hartley, Shannon and others. Thus in a standard text such as D. A. Bell's *Information Theory and Its Engineering Application* (1968) one finds stated clearly on p. 1: "Information ... is a measurable quantity which is independent of the physical medium by which it is converted. "This does not necessarily mean that it has a physical reality. Bell compares information to the more abstract term "pattern". However, it does imply an existence of its own.

Although the communications engineers treat information, follow this idea to its logical conclusion-viz, information exists. Perhaps part of the problem of recognizing and accepting the idea that information has physical reality and constitutes an intrinsic property of the universe, stems from the fact that we ourselves are so deeply embedded in the processing and transmitting of it.

(秦荻辉. 精选科技英语阅读教程[M]. 西安:西安电子科技大学出版社,2008.)

4.

The Chipping Norton Challenge for Driverless Cars

I am pleased that nurseries are considering the impact of materials they use in creative activities (A green guide to glitter alternatives, 20 November). The staff of the nursery school where I was headteacher for 10 years would be appalled at the suggestion that edible material such as cereals or pulses could be used as an alternative. We thought that allowing children to play with food that would be lifesaving for children suffering from malnutrition was a reinforcement of the superior attitudes that prevail in much of society.

(Elizabeth Martin, Bexleyheath, Kent)

The chancellor says the introduction of driverless cars will be very challenging (Driverless cars in four years' time, 24 November) and those who drive for a living will need retraining. The challenge will be to train the driverless delivery van bringing my parcels when I'm out to proceed up the drive, squeeze up a 2 ft-wide path, turn left to the side garage door and leave the goods on the bench. Oh, and to be careful reversing out.

(Margaret Bruce, Chipping Norton, Oxfordshire)

I couldn't agree more with Kate Phillips (Letters, 24 November). For us townies who don't often get to experience the full impact of rural life, a summertime car trip across the British countryside gives a shocking measure of the change of insect population, on our windscreens. Ten to 15 years ago I had to stop a couple of times on a three-hour journey to clean the windscreen just to see out. This year not a single insect over the same distance. No wonder there are hardly any spiders and birds left. Perhaps it was all my fault!

(Paul Huxley, London)

I live in Guildford, Surrey. Admittedly we've seen hardly any wasps and only had a few damselflies this year but we see plenty of small birds. For the last few days we've had a black cap feeding on the remaining apples; coal tits, long-tailed tits, great tits and of course blue tits on our feeders. Our record for goldfinches at once is 16 — the most common bird we get. The icing on the cake yesterday was a bullfinch.

(Linda Kendall, Guildford, Surrey)

There might be insects missing in other parts of Britain, but in May this year the fruit flies invaded my kitchen and, regardless of any kind of attack upon them, they stayed until about two weeks ago.

(Stuart Waterworth, Tavistock, Devon)

(https://www.theguardian.com/technology/2017/nov/24/the-chipping-norton-challenge-for-driverless-cars)

5.

On the arts scene, Lydia Cottrell — one-half of the live art duo, 70/30 Split (on Tumblr), who runthe Slap festival (Salacious Live Alternative Performance) — echoes this: "York is low on affordable space. In Leeds, you might get an empty shop unit to work in, but York's not like that. Property is at a premium and, as a city, it's very geared to tourists — not least its high-profile arts stuff. A group of us are looking at how we can enable smaller, higher-risk work, and, in particular, produce work for people who live in York. That's a very different thing."

It's a thing that, by using this guide on your next trip to York, you can actively support.

Music

The Crescent

A former working men's club and cabaret venue, the Crescent is now run by the people behind York'sIrie Vibes Sound System. It has retained its community spirit, its darts teams and billiard tables, but it is now a crucial creative fulcrum. This month saw Josie Long, Mr Scruff, Acid Mothers Temple, Mark Thomas and Peggy Seeger play there — an eclectic, international line-up of underground music and comedy, at a venue simultaneously focused on nurturing York talent.

A short walk from central York, this pub is renowned for everything from its street-food Sundays (last one on 19 November), to its natty beer garden. But primarily, it is a terrific music venue for both local and touring guitar bands and more experimental events. For instance, on 22 November, it hosts EMOM York, an open-mic night for electronic producers. "The owners, Chris Sherrington and Chris Tuke, are really serious about music," says Petersen, whose Bad Chapel party showcased Pye Corner Audio here this month. "One of them used to be a sound-engineer, and the pub has an amazing rig. It's a real audiophile's venue."

Mansion Underground

Every city needs one: a dark basement where, under its low vaulted ceilings, York's clubbers can lose themselves until the wee small hours to cutting-edge house, techno and bass music. Thepromoters BlackBox and Animaux, who welcome Untold on November 25, make particularly good use of Mansion's Funktion-One sound system.

Fibbers

This stalwart venue is still going strong after its move across town in 2014. Its program of, generally, mid-level touring bands and the occasional tribute act, tacks rather heavily to middle-aged nostalgia (Shed Seven, Big Country and Zodiac Mindwarp feature this winter), but from Lydia Lunch to New York "brasshouse" trio Too Many Zooz, there are edgier events happening here.

(https://www.theguardian.com/travel/2017/nov/15/alt-city-guide-york-food-music-art-theatre-drink-bars)

6.

Essaouira, Morocco

With an empty nest last year we decided we wanted sun and we wanted it as cheap as possible. So we took a £30 easy Jet flight from Luton, stayed at the beautiful Riad Dar Adul (from £49 a night for two) and took in all that this west coast town's medina had to offer — from the amazing fish restaurants to the artsy kasbah. Sitting on endless pale sand in 22℃ heat made for an unforgettable November.

Koh Lipe，Thailand

Baby-soft white sand, crystal-clear water, fresh fruit smoothie stands, and unlimited padthai! The tiny island of Koh Lipe is accessible only by boat and there are two mandatory arrival fees: a £1.20 taxi boat fee and a £4.50 national park fee. I enjoyed my stay at A-Plus Hotel for £15 a night. A-Plus provided a clean environment with strong air conditioning and powerful showers. After enjoying complimentary fresh fruit, I walked 200 metres to the coast where I could sunbathe, snorkel and dive. At night, cold Chang beer with stunning views of the Andaman Sea is a must!

Cassandra Tulloh

Stellenbosch，South Africa

Take advantage of the favorable exchange rate and soak up the sun in the vibrant university city of Stellenbosch, an hour's drive from Cape Town. Set among vineyards of the Cape Winelands, South Africa's second-oldest city has oak-shaded streets, wonderful arts and cafe culture and a diverse and welcoming populace. We used HomeAway. Its properties start at £23 a night in Jan/Feb. Stellenbosch is a foodie paradise with a wide spectrum of world-class eateries. Eat at community restaurant Twaalf, and as you savour delicious braai (barbecued meat) and other local delicacies, you will also be helping the homeless.

Los Cristianos, Tenerife

Forget everything you think you know about the southern coast of Tenerife and head to LosCristianos. You are pretty much guaranteed lots of beautiful sunny days and the sand is amazing. There are lots of gems to be found among the tourist haunts, such as Restaurant El Cine which is a classic, and there are still little "hole in the wall" bars where you can grab a cortado and a tapa with the locals. And who can resist a walk along the beach on Christmas Day?

Jody Levitus

Goa, India

We've just got back from an amazing holiday in Goa, staying in the far south at The Cape Goa. While some parts of Goa are getting a bit trashed nowadays, this little hidden corner in the south still holds on to that fabled oldGoan charm. Our beach hut had the most amazing views of sunset over its own private beach. We hired mopeds and enjoyed nights out in Palolem and lazy days exploring Agonda and Cola Beach. OK, so it's not the cheapest beach bungalow in India (from £140 a night for two; we also stayed on Agonda beach for £4 a night), but it's definitely the most luxurious. It has a jacuzzi! Make sure you ride to Betul Fort and the Sal estuary, past laid-back villages and old colonial Portuguese houses. Bliss...

(https://www.theguardian.com/travel/2017/nov/23/winter-sun-getaways-readers-tips-algarve-thailand-caribbean)

7.

Shifting Attitudes Among Democrats Have Big Implications for 2020

Sen. Bernie Sanders (I-Vt.) speaks during a health-care rally at the 2017 Convention of the California Nurses Association/National Nurses Organizing Committee on Sept. 22 in San Francisco. Sanders is pushing his "Medicare for All" bill in the Senate. (Justin Sullivan/Gettty Images)

By Dan Balz Chief correspondent October 7 at 10:05 AM

Partisan divisions are not new news in American politics, nor is the assertion that one cause of the deepening polarization has been a demonstrable rightward shift among Republicans. But a more recent leftward movement in attitudes among Democrats also is notable and has obvious implications as the party looks toward 2020.

Here is some context. In 2008, not one of the major candidates for the Democratic presidential nomination advocated legalizing same-sex marriage. By 2016, not one of those who sought the nomination opposed such unions, and not just because of the Supreme Court's rulings. Changing attitudes among all voters, and especially Democratic voters, made support for same-sex marriage an article of faith for anyone seeking to lead the party.

Trade policy is another case study. Over many years, Democrats have been divided on the merits of multilateral free-trade agreements. In 1992, Bill Clinton strongly supported the North American Free Trade Agreement (NAFTA) in the face of stiff opposition from labor unions and others. He took his case into union halls, and while he didn't convert his opponents, he prospered politically in the face of that opposition.

By 2016, with skepticism rising more generally about trade and globalization, Hillary Clinton was not willing to make a similar defense of the merits of free-trade agreements. With Sen. Bernie Sanders (I-Vt.) bashing *the Trans-Pacific Partnership* (TPP) as a presidential candidate, Clinton joined the chorus of opponents. She ended up on the opposite side of then-President Barack Obama, even though she had spoken warmly about the prospects of such a treaty as secretary of state.

Looking ahead to 2020, something similar is likely to take place on the issue of health care. Because of changing attitudes that already are underway within the party, it will be difficult for any Democrat seeking the nomination not to support some kind of single-payer health-care plan, even if big questions remain about how it could be accomplished.

Sanders used his 2016 presidential campaign to advocate a universal health-care plan that he dubbed "Medicare for All." The more cautious Clinton, who saw flaws in what Sanders was advocating, argued instead for focusing on improvements to *the Affordable Care Act*.

Sanders has now introduced a "Medicare for All" measure in the Senate, and his co-sponsors include several other prospective candidates for the Democratic nomination in 2020.

Meanwhile, a majority of House Democrats have signed onto a single-payer plan sponsored by Rep. John Conyers Jr. (D-Mich.) that goes much further. This has happened even though some of those who like Conyers's idea in principle question whether it is ready for prime time, not only because of the potential cost and the absence of a mechanism to pay for it, but also because of other potential policy flaws as well.

The pressure to embrace single-payer plans grows out of shifts in attitudes among Democrats. The Pew Research Center found in June that 52 percent of self-identified

Democrats now support a government-run health-care system. That is up nine points since the beginning of the year and 19 points since 2014. Among liberal Democrats, 64 percent support such a plan (up 13 points just this year) and among younger Democrats, 66 percent say they support it.

Health care isn't the only area in which Democratic attitudes are shifting significantly. Others include such issues as the role of government and the social safety net; the role of race and racial discrimination in society; and immigration and the value of diversity.

A few days ago, the Pew Center released a comprehensive survey on the widening gap between Republicans and Democrats. The bottom line is summed up by one of the opening sentences in the report: "Republicans and Democrats are now further apart ideologically than at any point in more than two decades."

This poll is the latest in a series of surveys dating to 1994. Together they provide not just snapshots in time, but also an arc of the changes in public opinion. Republicans moved to the right harder and earlier than Democrats began moving left, and their base remains more uncompromising. But on a number or questions, the biggest recent movement has been among Democrats.

In its new survey, Pew found the widest partisan gap ever on the question of whether government should help those in need — primarily because of recent shifts among Democrats. From 2011 to today, the percentage of Democrats who say government should do more to help those in need has jumped from 54 percent to 71 percent.

Only a minority of Republicans (24 percent) say government should do more for the needy, and that figure has barely moved in the past six years. The Republicans shifted their views from 2007 through 2011, the early years of the Obama presidency, during which their support for a government role dropped by 20 percentage points.

Two related questions produce a similar pattern among Democrats. Three in 4 Democrats say that "poor people have hard lives because government benefits don't go far enough to help them live decently," up a dozen points in the past few years.

Eight in 10 Democrats say the country needs to continue to make changes to give blacks equal rights with whites, up 18 points since 2014. And more than 6 in 10 say "racial discrimination is the main reason many black people can't get ahead these days," up from 4 in 10 three years ago.

Meanwhile, only a quarter of Republicans agree with the statement on government benefits, fewer than 4 in 10 say the country needs to continue to do things to provide equal rights for blacks, and just 14 percent cite racial discrimination as the main reason many blacks can't get ahead.

Members of both parties have become more positive in their attitudes about immigration in recent years, but the partisan gap remains huge — 42 points in the new survey. Today, 84 percent of Democrats say immigrants strengthen the country through hard work and talents, up from 48 percent in 2010. In 2010, 29 percent of Republicans agreed with that statement; today, that's risen to 42 percent.

Why have Democratic positions moved so dramatically and so recently on these questions race and government and immigration? Though it is not explicitly addressed in the survey, one possible reason is a reaction to the 2016 campaign and the Trump presidency.

President Trump obviously found strong support for his controversial views on immigration, whether his call to build a wall on the U.S.-Mexican border or to bar refugees from countries mostly in the Middle East. Those pronouncements helped him win the presidency. But those policies and the rhetoric that often preceded them also produced a strong backlash from the president's opponents.

The 2016 campaign ended up highlighting issues of national identity — race and immigration and the shifting character and face of the country — in often divisive ways that unleashed the kind of ugliness seen in Charlottesville in August.

The Democratic Party is being shaped by the Trump presidency and by reactions to the president among rank-and-file Democrats. Party leaders have been taking notice since Trump was sworn in as president and have moved as well.

Those who seek the party's nomination in 2020 understandably will be guided by these sentiments. But they must find a way to harness the movement into a political vision that is attractive to voters beyond the Democratic base — a vision that is grounded not just in anti-Trump resentment but in fresh and sound policies as well. In such polarizedtimes, that will not be easy.

(https://www.washingtonpost.com/politics/shifting-attitudes-among-democrats-have-big-implications-for-2020/2017/10/07/a1741398-aae1-11e7-850e-2bdd1236be5d_story.html? hpid=hp_hp-more-top-stories_sundaytake-1008am%3Ahomepage%2Fstory&utm_term=.810271147961)

8.
United Airlines CEO Offers Softer Apology Following Stock Nosedives

The CEO of United Airlines has issued a second public apology about the man who was forcibly removed from a flight on Sunday, calling the incident "truly horrific".

"No one should ever be mistreated this way," Oscar Munoz wrote in a note to employees Tuesday, one day after video posted by fellow passengers showing police dragging the man off the plane went viral.

Munoz was criticized after his official statement on Monday described the violent removal as an effort to "re-accommodate" passengers. He also described the man as "disruptive and belligerent". As the company's share prices plunged on Tuesday, however, the executive turned attention back on to the company.

Nearly $1bn of the company's value was erased in trading on Tuesday. Later that day, Munoz said he was committed to "fix what's broken so this never happens again". He pledged to reviewthe company's policies for seeking volunteers to give up their seats, for handling oversold flights and for partnering with airport authorities and local law enforcement.

The company plans to share results of the review by 30 April.

The value of the carrier's holding company, United Continental Holdings, had fallen over 4% before noon, knocking almost a billion dollars off its value. It rallied slightly, leaving the share price down 2.8%, close to $600m less than the company's $22.5bn value as of Monday's close.

Investors had largely shrugged off United's woes during trading on Monday. The airline's stock finished Monday's trading session 0.9% higher, adding about $200m to the company's market cap.

But the airline's problems only seem to have escalated since Sunday, when a man was violently removed from a flight by aviation police officials at Chicago's O'Hare international airport after refusing to volunteer his seat on the overbooked flight.

In one video clip, guards aggressively grab then drag the passenger down the aisle of the plane as other passengers shout: "Oh my God" and "Look what you did to him".

Sean Spicer, the White House press secretary, said he was "sure" that Donald Trump had seen the video.

"I don't think anyone looks at that video and isn't a little disturbed that another human being is treated that way," he told reporters. "Clearly watching another human being dragged down an aisle, watching blood come from their face after hitting an armrest or whatever, I don't think there's a circumstance that you can't sit back and say this probably could have been handled a little bit better, when you're talking about another human being."

Spicer noted that United was conducting a review and described it as a local matter that did not necessarily require a federal response. "I think there's plenty of law enforcement to review a situation like that and I know United Airlines has stated that they are currently reviewing their own policies. Let's not get ahead of where that review goes."

On Tuesday, Munoz attracted criticism after calling the passenger "disruptive and belligerent" and saying that the airline's employees had "followed established procedures".

The passenger is overheard in one video claiming he was being profiled for being Chinese, and the video has caused outrage in China and calls for a boycott.

Ahead of the market's open, United shares had been down by as much as 6% in premarket trading.

(https://www.theguardian.com/us-news/2017/apr/11/united-airlines-shares-plummet-passenger-removal-controversy)

9.

Telemedicine holds great promise within mainstream health care. Countless trials are under way to assess technology that can monitor people who have been diagnosed with heart conditions, or diseases like diabetes, from the comfort of their own homes. Rather than having their devices periodically checked at a clinic, some pacemaker patients can now have their implants inspected via mobile phone. That way, they need only visit the clinic when it is absolutely necessary.

Similarly, BodyTel, based in Germany, is one of several firms to have developed sensors based on Bluetooth wireless technology that can measure glucose levels, blood pressure and weight, and upload the data to a secure web server. Patients can then manage and monitor their conditions, even as they give updates to their doctors. Honeywell, an American industrial giant, has devised a system that patients can use at home to measure peak flow from their lungs, ECG, oxygen saturation and blood pressure, in order to monitor conditions ranging from lung disease to congestive heart failure. Doctors continually review the data and can act, by changing the patients' medication, for example, if they spot any problems.

This sort of thing appeals to both patients and health-care providers alike. The patients keep their independence and get to stay at home, and it costs less to treat them. And as populations age in developed countries, the prospect of being able to save money by treating people at home looks increasingly attractive.

It is not just people with diagnosed conditions who are starting to receive this kind of equipment. Since 2006, Britain has spent £80m ($160 million) on "preventative technology grants" which provide special equipment to enable 160,000 elderly people to stay in their homes.

Most of today's technology, however, calls on the patients to remember to monitor themselves, and also requires them to operate the equipment. For some patients, such as those in the early stages of Alzheimer's disease, that is impractical. So a lot of work is being done to automate the monitoring process and make the equipment easier to use.

William Kaiser and his colleagues at the University of California, Los Angeles, have developed a "smart cane" to help monitor and advise people convalescing at home, for example. "It has force sensors that measure pressure at the tip of the cane and around the handle. It also has motion sensors and accelerometers," says Dr. Kaiser. It uses these to calculate the gait of the patient and work out how they are doing with the cane, giving them feedback about how they could make better use of it to recover from, for example, a hip replacement. "1t provides guidance, either as beeps or it can talk to you," he says.

Another approach is to use sensors embedded in the home. Oliver Goh of Implenia, a Swiss building-management firm, has come up with a system to monitor the well-being of the occupant of a house. Using sensors on doors and mattresses, smart pill boxes that can tell when they are being opened, heart-monitors and a location-sensing wristwatch — the system allows carers to keep tabs on elderly people. "Implenia now has six elderly volunteers lined up to test the technology," says Mr. Goh. He hopes that if they have a heart attack, cannot get out of bed or need help, their carers will soon know. "Ultimately," he says, "the aim is to see if this sort of approach can help to extend life expectancy."

Prevention is better than cure.

Looking even further ahead, some day it may make sense to give these technologies to healthy people, the "walking well". If sensors can monitor people without a threat to their privacy or comfort, doctors may able to spot diseases before the patient notices any symptoms. "It's moving from telemedicine to telehealth andteleprevention," says Dr. Grundy of IBM. It could also improve the efficiency of health-care systems, he says.

This kind of approach could save money as well as spotting illnesses early, says Dr. Kaiser. "We'll detect them earlier when the cost of treatment and impact on an individual will be less," he says. The technology for this does not yet exist, admits John Linkous, executive director of the American Telemedicine Association. "There still isn't a device that can give you a complete body check," he says. "But I'm very optimistic about it in the long run."

One idea is to use wireless infra-red skin sensors to measure blood-count, heart rate and the level of oxygen in the blood. Another is to implant wireless sensors powered by the wearer's own body heat. Yet another common idea is to use smart toilets that can monitor human waste for the telltale signs of intestinal disease or cancer. The hard part is not so much developing the sensor technology, says Dr. Linkous, as sifting through the results. "It would produce a tsunami of data, and the problem is that we aren't set up with health-care systems that can deal with all that," he says.

The answer will be even more technology, says Dr. Bakalar. "There has to be a way of filtering this information so that it doesn't overwhelm the medical services," he says. The obvious approach is to use "expert systems"— software programmed with expert medical knowledge and that can make clinical judgments.

Like telemedicine, expert systems have been around for some time. Trials in Denmark, to advise doctors how to prescribe, suggest the technology has great scope. Sometimes they can reach better clinical judgments than human experts do. But they are not widely used, partly because doctors are unwilling to be bossed around by a computer in the corner, but also because they have been difficult to integrate into medical practice. They could be ideally suited to telehealth, however, quietly sifting through the data generated by sensors and only raising the alarm and calling in their human colleagues when it becomes necessary to do so.

The shift from telemedicine to telehealth reflects a broader shift from diagnosis and treatment to "wellness". Taken to its technological conclusion, this would involve using wireless sensors and implants to screen entire populations for early signs of disease as they go about their daily lives. If it can be made to work, the days of making an appointment to see your doctor when you are not feeling well could be over. Instead, it may be your doctor who calls you.

(韩孟奇. 科技英语阅读[M]. 上海：上海交通大学出版社，2012.)

10.

Property Unaffordable for 100,000 Households a Year in England

Report finds number of households priced out is up by 30,000 a year owing to shortage of affordable homes to rent or buy.

The autumn statement reiterated a government pledge to build 300,000 homes a year. Photograph:Dinendra Haria/Rex/Shutter stock

Almost 100,000 households in England are being priced out of the property market each year because of a shortage of affordable homes to rent or buy, according to a report.

Research by the estate agent Savills, shared with *the Guardian*, found the number of priced-out households had risen from its previous projection in 2015 of 70,000 a year. This was in part because of a change in how housing need is assessed, but also due to rising prices and stagnant wage growth.

On Wednesday, the chancellor, Philip Hammond, reiterated a pledge to build 300,000 homes a year by the mid-2020s and unveiled a package of measures to support housebuilding. Savills said one-third of those would need to be offered at below market prices to meet the growing need for affordable homes.

About 96,000 households are unable to afford homes at the market rate, either to buy or rent, Savills said, with the vast majority in London and the south. It said varying approaches were needed to address the shortfall in different parts of the country: low-cost rented homes were needed more in markets in which incomes were smaller, while a mixture of homes, including shared ownership, would help in more expensive areas.

In London, 20% of the households affected have incomes above £35,000, Savills said, while the same proportion earn less than £10,000.

Over the past three years, 55,000 fewer affordable homes have been built each year than were needed, the research found. Although 42,500 households in the capital required below market rate housing, only 8,800 affordable homes a year had been delivered. In the south outside London, 15,500 affordable homes a year were being built while 34,100 were needed.

Meanwhile, in the north of England, low incomes were locking out 9,600 households a year, with 8,900 homes being delivered.

Although the post-budget headlines were dominated by Hammond's decision to scrap stamp duty for the majority of first-time buyers, his announcement included more money for building and rule changes to help councils provide housing.

A speech by the prime minister, Theresa May, at the Conservative party conference in October made a commitment of £2bn over four years to fund social housing. However, to house 100,000 emerging households over this period would need funding of £7bn a year, Savills said.

Paying for the new homes would reduce the housing benefit bill by £430m a year.

Helen Collins, the head of housing consultancy at Savills, said the budget had offered many positives for the housing sector, but expressed some concerns.

"While the chancellor had plenty of good news, we feel there are still some important issues missing from housing policy," she said.

These included no changes on building on greenbelt land and no additional funding for homes for social rent. "There was little reference to the importance of social rented homes as a better and cheaper alternative to the private rented sector for very low-income households," Collins said.

(https://www.theguardian.com/business/2017/nov/27/property-england-priced-out-households-affordable-homes-savills-report)

二、汉译英

1.

麒麟花

（文章来源：华南植物园；发布时间：2014-10-23）

麒麟花(Euphorbia splendens Bojer ex Hook.)别名铁海棠、虎刺梅、番仔刺。大戟科(Euphorbiaceae)落叶灌木,原产非洲马达加斯加群岛。它的茎上密生褐色的硬刺,花柄长而挺出,双双成对排列,花形小,无花瓣,肾状而鲜红的部分是苞片,因其颜色鲜红而常被人们误认为花瓣。其花期很长,从3月一直开到12月,但以春季开花最佳,因为它的花型似梅,茎上有硬刺,故有"虎刺梅"之称。它的寿命较长,盆栽能活几十年,因为生长缓慢,可扎成各种形态的盆景。它的花在阳光照射下,鲜艳夺目,经久不凋,是很好的室内装饰盆花。

(http://www.cas.cn/kxcb/kpwz/201410/t20141024_4230401.shtml)

2.

摘要：本文系新近由高等教育出版社出版的《英汉语篇翻译教程》的导论。导论在衔接理论的完整框架的范围内表述了我们的理论指导实践、实践印证和充实理论的翻译思想。本文论及语法上的衔接,涉及照应、替代、省略、连接四个方面。文中实例均取自该教材。

(郭立秋,范守义,贾令仪.语篇翻译中的衔接问题:理论解读与翻译应用[J].上海翻译,2011(4):31-34)

3.

2014年贵州省关于进一步完善药品生产监管机制的意见

为了进一步完善药品生产监督管理机制,明确工作职责,提高工作效率,方便企业办事,根据《国务院关于取消和下放一批行政审批项目等事项的决定》(国发〔2013〕19号)、《麻醉药品和精神药品管理条例》(国务院令第442号)、《药品生产监督管理办法》(2004年局令第14号)精神和《关于印发国家食品药品监督管理总局2014年工作要点的通知》(食药监〔2014〕7号)要求,经研究决定,对完善药品生产监管工作机制提出以下工作意见:

一、药品生产企业在筹建申报时,应先报市、州食品药品监督管理局初审并进行现场查看,在符合当地政府产业发展规划和有关申报条件后,出据审查意见由企业一并报省局审批。

二、药品生产企业在申请GMP认证检查时,市、州食品药品监督管理局要先进行初审和现场初查,并出据明确的意见后由企业一并报省食品药品监督管理局审查。

三、药品生产企业质量负责人、生产负责人发生变更的,企业在变更后10个工作日内将变更人员简历及学历证明等有关情况报所在地市、州食品药品监督管理局登记,市、州食品药品监督管理局在5个工作日内传真到省食品药品监督管理局备案。

四、药品生产企业质量授权人发生变更时,企业应先将变更人员简历及学历证明等有关情况报所在地市、州食品药品监督管理局进行初审后再报省食品药品监督管理局备案。

五、药品生产企业的关键生产设施等条件与状况发生变化的,应当在发生变化20个工作日内报所在地市、州食品药品监督管理局登记,企业在10个工作日内报省食品药品监督管理局审查,省局根据需要可委托市、州食品药品监督管理局进行检查,符合《药品生产质量管理规范》要求的省局进行备案。

六、麻醉药品、精神药品和药品类易制毒化学品生产普通制剂和定点生产二类精神药品的药品生产企业应当每月向所在辖区市(州)食品药品监督管理局书面报告本企业特殊药品的购进、使用和销售情况。市(州)食品药品监督管理局应当每3个月向省食品药品监督管理局报告本辖区特殊药品监管的相关情况,实行特殊药品监管工作向局领导的季报制。

七、药品生产企业供应商发生变更的,企业在变更后15个工作日内将变更情况报市、州食品药品监督管理局备案。

八、药品委托生产和委托提取加工的,药品生产企业委托方应报企业所在地市、州食品药品监督管理局初审后报省局审批。委托生产日常监管由受托方所在地食品药品监督管理局进行监管。

九、以上完善后的工作机制从2014年4月起开始执行。

(https://wenku.baidu.com/view/ba0ccc4deefdc8d376ee32a4.html)

4.

推荐信

四川大学经济学院财政专业06级本科生×××是本人《××××》课班上的学生。该生在学习该课程的过程中积极主动,认真踏实,并在本课程取得了良好成绩,在专业名列前茅。

在教授该课程时,我比较注重对学生运用基本理论和基础知识分析和思考问题的能力及实际计算和操作能力的培养。该生在学习期间表现比较突出,不仅上课认真听讲,能提出很有独创性的问题,而且对我提出的案例分析题,勇于上台发言,能清楚地阐述自己的观点。思维活跃,思考问题思维活跃,很能深入到问题的本质,具有较强的分析问题的能力;专业基础扎实,能灵活运用所学的基本理论和基础知识。

该生综合表现突出,予以推荐,希望审核通过。

推荐人:×××

(https://wenku.baidu.com/view/f920dc6f27d3240c8447ef65.html)

附录2 翻译补充练习参考答案

一、英译汉

1.

【参考译文】

人体克隆

凯文 · 邦萨尔(Kevin Bonsor)

韩国科学家声称已通过克隆技术成功创造出人类胚胎。治疗性克隆技术可用于治疗帕金森病和阿尔茨海默病等顽疾,这可能会引起医疗领域的根本性变革。

1997年,苏格兰一群科学家创造出"多莉"克隆羊。整个世界为之哗然。很多人认为在未来10年内,我们将听到更令人震惊的消息——首位克隆人诞生。韩国科学家已利用克隆技术创造出人类胚胎干细胞。

迄今为止,人体克隆还仅存于电影特效中,但科学的自然发展使人体克隆已成为真正可能实现的技术。既然我们已克隆出绵羊、老鼠和牛,为何不能克隆人?尽管部分国家已制定禁止人体克隆的法律,但在其他许多国家,人体克隆仍然合法。

我、我自己和我的克隆体

2001年1月,由肯塔基大学前任教授帕纳约蒂斯·扎沃斯(Panayiotis Zavos)和意大利研究人员塞维利诺·安迪诺瑞(Severino Antinori)领导的科学家团队声称,该团队计划在未来两年内克隆人体。大约在同一时间,《纽约邮报》报道称,一对美国夫妇计划向拉斯维加斯市克劳内德公司(Clonaid)支付50万美元,用以克隆他们幼年夭折的女儿。

克隆领域的科学家以科学的名义可能在追求成功完成首例人体克隆的荣耀。无论科学家动机如何,我们都很有可能会在下一个10年中的某个晚间新闻中看到成功克隆第一例人类婴儿的报道。科学家指出,当前的技术可用于克隆动物,但通过分娩创造克隆胚胎的成功率仍非常低。

如果实施人体克隆,科学家可采取的一个方法是体细胞核移植。该方法与创造"多莉"

羊的克隆程序相同。首先在体细胞移植阶段,科学家需从供体中取出卵子,再通过去除卵子核创建无核卵子;然后从克隆人体获得含有 DNA 的细胞;接着,使用电流将去核卵子和克隆对象的细胞融为一体。这样即创造出一个胚胎,通过体外受精植入代孕母体。如果该程序进展顺利,则代孕母亲在正常妊娠期结束时产下克隆婴儿。当然,该程序成功率或许仅有 1‰ 至 2‰(科学家通过 277 次尝试才创造出"多莉"羊)。下图展示的是体细胞移植克隆的流程。

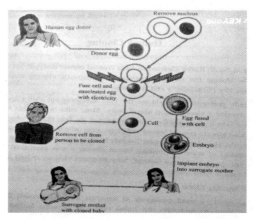

一些科学家似乎认为人体克隆是不可避免的结果,但是我们克隆人体的目的何在?当前有诸多原因可能致使人们转向支持人体克隆。我们现在探讨其中的一些原因。

谁将需要克隆技术?

并非所有的人体克隆都涉及创造完整的新人体。人们认为,克隆可能会帮助患有严重疾病的患者康复。克隆技术的一项潜在应用是创建人体修复包。换句话说,科学家可以通过克隆细胞,修复导致疾病产生的突变基因。2001 年 1 月,英国政府颁布法规,允许克隆人体胚胎,治疗帕金森病和阿尔茨海默病等疾病。

尽管民众完全接受人体克隆还需要漫长的时间,但治疗性克隆技术已迈出第一步。治疗性克隆是指利用人体 DNA 培养克隆胚胎的技术。然而,应用克隆技术时,无需将克隆胚胎植入代孕母体,而是利用克隆胚胎细胞培养干细胞。经克隆后培养的干细胞可视为人体修复包。利用干细胞可以培养替代器官(如心脏、肝脏和皮肤等)。此外,干细胞还可用来培养神经元,以便治疗阿尔茨海默病、帕金森病或雷特综合征患者。

治疗性克隆工作流程如下:

- 从患者身上提取 DNA。
- 然后,将 DNA 注入无核供体卵中。
- 然后,像典型的受精卵一样,该卵分裂,形成胚胎。
- 从胚胎中取出干细胞。
- 利用干细胞可培养出治疗患者所需的任何类型组织或器官。

还有人认为,克隆技术可向患有不育症夫妇提供帮助,这样出生的孩子至少有父母中一个人的生物学属性。扎沃斯和安迪诺瑞声称,为不孕症夫妇提供帮助是其研究目标。扎沃斯指出,已有数百对夫妇排队等待支付约 5 万美元,以获得这项服务。该团队声称,克隆程序具体如下:从不孕男子身体获取细胞,将该细胞注入卵子,然后将卵子植入女子的子宫内。

这样,他们的孩子长相将与父亲相同。

人体克隆技术的另一用途是使已故亲人重返人间。我们可以想象一下利用你曾祖母的DNA片段创建曾祖母克隆体的场景。从某种意义上说,你将成为你曾祖母的父母。这种做法将打开伦理的潘多拉魔盒。这一魔盒可能很快就会打开。一对美国夫妇向克劳内德公司正在支付50万美元,以利用保存的皮肤细胞克隆其已故的女儿。

到底该不该克隆人体?

对于克隆技术这一存在争议的科学领域,批评者经常质疑:"我们有能力进行人体克隆,难道就意味着我们应该克隆人体吗?"随着人体克隆技术日益成熟,关于人体克隆的争议也越来越激烈。人体克隆反对者声称,尽管克隆给我们带来益处,但也会带来伤害。另外一个争议的焦点是,如何规范克隆程序。

尽管当前美国联邦政府尚未制定禁止人体克隆的法律,但已有若干个州通过颁布州政府法律禁止克隆人体。美国食品药品管理局(FDA)也指出,在美国试图实施人体克隆的任何人员都必须首先获得该管理局批准。在日本,实施人体克隆是犯罪行为,可判处最高10年监禁。尽管英国政府已允许克隆人体胚胎,但政府也同样在通过立法努力阻止克隆完整人体。

尽管法律可对人体克隆造成威慑作用,但一些科学家认为,当前的克隆技术距离人体克隆还很远。"多莉"羊的共同创造者之一伊恩甚至声称,人体克隆是不负责任的犯罪行为。克隆技术仍处于早期阶段,大约98%的克隆尝试以失败告终。(克隆出的)胚胎或者不适合植入子宫,或者克隆体在妊娠期间或在出生后不久死亡。

即便是存活下来的克隆体也可能出现遗传基因异常状况。一些克隆体存在心脏或肺部疾病、糖尿病、血管疾病或免疫系统功能性障碍。其中较知名的案例是,一只出生后的克隆羊,因动脉畸形导致其患有肺部慢性过度换气症。

人体克隆反对者指出,尽管我们可以对存在缺陷的克隆动物采取安乐死做法,但是如果在人体克隆过程中发生类似状况,那将产生道德问题。人体克隆支持者认为,在将克隆胚胎植入母体子宫之前,现在的技术已经能够较为容易地挑选出有缺陷的克隆胚胎。当前关于人体克隆的争议才刚刚开始,随着科学的不断进步,人体克隆可能会成为21世纪最大的伦理难题。

2.

【参考译文】

语料库语用学中的言语行为:自发性和诱发性话语中指示词的定量对比研究

Ilka Flöck 和 Ronald Geluykens

摘要

本项研究旨在比较不同条件下收集的三种不同语言语料指示词:①自发性口语数据(摘自英语国际语料库中的英国部分);②自发性书面语数据(即商业信函);以及③诱发性书面数据(通过话语完成任务收集)。研究结果表明,自发性和诱发性数据组之间以及口头和书面数据之间存在显著差异。此类差异既表现在所谓的指令性中心语中,也出现在伴随中心语的调整策略降级和升级中。该状况致使在所有三个数据组的指令实现方面,均存在不同程度的直接性。这些结果显示基于真实言语语料数据收集方法和话语类型方面定量比较研究的重要性。

3.

【参考译文】

人类信息的存在及其传播方式

正如存在不同形式的能量（如机械能、化学能、电能、热能、声能、光能和核能等），也存在不同形式的信息。人类信息仅代表信息的一种形式。稍后我们会讨论非人类信息的形式。然而，人类信息本身有各种不同的存储和传播方式，代表很多不同的形式。

存储和处理人脑中信息的系统极其复杂和神秘，构成了生物科学研究最新的前沿领域。与计算机相比，人脑至少在三个方面表现出更复杂的特性（如斯通尼尔1984年发表评论所述）。首先，人脑回路复杂得多：人脑不仅包含10个细胞顺序，而且每个脑细胞向外还可能与数以千计的其他脑细胞相连接。每个神经元的功能类似于晶片机而非晶体管。其次，传输系统不同。在计算机中，传输系统主要处理沿着导体移动的电子。与之相反，神经脉冲涉及细胞膜的渐进去极化过程。神经脉冲可以更加微妙地调节传输系统，与第三个主要差异有关：当代计算机处理的信息为数字化信息，而人类神经系统中存在数十种可增强或抑制神经脉冲的神经递质和其他相关物质，即人类神经系统是一个具有精密调节和集成功能的模拟设备网络。

人脑中的信息与计算机信息存在本质差异。此外，两人之间、两台计算机之间或人与计算机之间信息传播形式同样存在差异。正如存在多种形式的信息，同样存在各种不同的信息传输或转换（由一种形式转换为另一种形式）方式。

看看本页面中的信息吧，这些信息正通过光线传递至读者的眼睛。撞击视网膜的光线通过细胞膜有序去极化，再转换为神经脉冲。在人脑神经元之间的突触中，信息被转换为化学神经递质脉冲，而这一过程逆向触发进一步的神经活动，向多个方向分化。最终，上述转换过程产生了大量的人脑活动：短期记忆存储在多个层面上与现有信息比较（从比较印刷字母和单词的图片及其含义，到比较本文的概念和读者对世界的看法等）、长期记忆存储，以及其他大量与吸收和分析新信息相关、仍然处于神秘状态的思维过程。在未来的某个时候，读者可能会将神经脉冲转换到声带上，把大脑中的神经信息模式转换为声波。声波代表信息的机械编码。声波冲击听者的耳朵，内耳中的纤维毛发状细胞器通过运动，可以把信息从机械脉冲转化为神经脉冲。这些神经脉冲进入听者大脑之后，信息在听者大脑中的处理过程与最初在读者大脑中的处理过程相类似。

或者，读者可以通过电话交谈。这时，信息从以声速传输的压缩空气分子模式转换到沿着铜导线以近似光速传播的电子脉冲模式，而电子脉冲继而可能被转换为沿着光纤传输的光脉冲。另外，读者可能会对着麦克风讲话。这时，信息被转换为穿越电波的电磁波模式。读者也可以用录音机录音，这时电磁脉冲转换为磁脉冲，通过磁带中的原子响应磁性，将磁脉冲"冻结"在磁带上，通过物理作用在磁带中按序排列成信息模式。

上述过程代表人类信息传播的一个完整周期。请注意该周期包含的信息传播形式如下：

1. 光线模式（从书本到眼睛）。
2. 细胞膜去极化脉冲（从眼睛到大脑）。

3. 化学物质脉冲(单个神经之间)。

4. 压缩空气分子脉冲,即声波(由发言者喉部发出)。

5. 液体或固体中的机械变形脉冲(在耳朵或送话器中)。

6. 电话线中的电子脉冲。

7. 光纤中的光脉冲。

8. 无线电波脉冲。

9. 磁脉冲(在电话听筒或无线扬声器中)。

这些信息在书本(即印刷体)、人脑和磁带中存储下来。

其中,第一种涉及染料分子模式,第二种涉及神经连接模式,而最后一种涉及磁化区域模式。在人眼视网膜、神经细胞之间的突触、喉部、内耳、电话、无线电发射器和磁带录音机中,信息从一种形式转换为另一种形式。

读者也可以决定信息的存储方式,通过影印或打字将信息存在电脑或文件中,从而再次将信息转化为叠加在纸分子上的染料分子模式。从洞穴壁画、木雕或石雕,到磁泡记忆和卫星通信,人类信息可以按照很多不同方式存储、传输或转换。这些方式仍在不断增加。

需要注意的是,信息传播方式(如上述九种方式)通常涉及波脉冲(光波、声波、无线电波)、电子脉冲或影响物质及其组织的脉冲。实际上,通信工程师将信息分成离散数据包打包传输,便于多人同时使用。根据哈特利(Hartley)、莎伦(Shannon)和其他先驱者的开创性著作所述,通信工程师自然明白信息是一个独立实体。因此,在贝尔(D. A. Bell)著的《信息理论及其工程应用》(1968年出版)(标准文本之一)的第一页写到:"信息是独立于所转换物理介质的可测变量",但这并不意味着信息具有物理现实性。贝尔把信息比作更抽象的术语"模式",但信息本身暗示其存在性。

尽管通信工程师处理信息,但顺着这一思维得出了合乎逻辑的结论,即信息具有存在性。我们承认和接受信息具有物理现实性以及信息构成宇宙固有属性这样的观点,或许问题出在我们"从处理和传输角度看待信息"这一做法上。

4.

【参考译文】

奇平·诺顿无人驾驶汽车面临的挑战

对于托儿所在创意活动中考虑使用相应材料产生的影响,我表示非常欣慰(11月20日发布了亮片替代品的绿色指南)。我曾在托儿所担任所长10年。我的员工对使用谷物或豆类等作为亮片替代品的建议感到震惊。我们认为对于营养不良的孩子而言,允许他们把玩由救命食物制成的玩具将会强化社会中普遍存在的优越感。

(伊丽莎白·马丁,肯特郡贝克里斯黑斯市)

英国财政大臣声称,推出无人驾驶汽车不仅是一项极具挑战性的任务(四年内推出无人驾驶汽车,11月24日),那些靠开车谋生的司机将接受职业再培训。这一挑战是:训练为我运送包裹的无人驾驶运输车,在我出门在外时,挤上仅2英尺宽的道路上,向车库侧门处左转,并把包裹放在长条凳上。另外,无人驾驶运输车倒车时还需要特别小心。

(玛格丽特·布鲁斯,牛津郡奇平·诺顿市)

我完全不赞同凯特·菲利普斯的观点(如11月24日信函所述)。对于我们这些不能经常体验农村生活的城镇居民而言,夏季英国乡村汽车之旅是我们非常乐意参与的活动,但在该旅程中,汽车挡风玻璃上昆虫种群的变化让我们感到惊讶。10至15年前,在3小时的旅程中,我得多次停车清理挡风玻璃上的昆虫,才能看清路况,但在今年同样距离的旅程中,挡风玻璃上竟然没有一只昆虫。难怪几乎没有任何蜘蛛和鸟类了。这可能都是我的错!

(保罗·赫胥黎,伦敦市)

　　我在萨里郡吉尔福德市居住。我得承认,我们今年几乎看不到任何黄蜂,只看到为数不多的豆娘蜻蜓,但是我们看到了很多小鸟。在过去几天里,大群小鸟(包括煤山雀、长尾山雀、大山雀和蓝山雀)啄食树上剩余的苹果,它们也在喂食器上啄食。我们的记录是,曾经一次性看到16只金翅雀(我们最常见到的鸟类)落下啄食。昨天,我们惊喜地见到一只红腹灰雀。

(琳达·肯德尔,萨里郡吉尔福德市)

　　在英国其他地方或许昆虫在逐渐消失,但今年5月果蝇却侵袭了我的厨房。无论我采用哪种驱虫措施,直到约两周前果蝇才消失。

(斯图尔特·沃特沃斯,德文郡塔维斯托克市)

5.

【参考译文】

　　在艺术方面,曾主办掌击节(the Slap festival)(现场另类表演)的现场二重奏之一、艺术家琳达·科特雷尔(Lydia Cottrell)(微博名为70/30 Split)回应道:"约克市可供使用的空间小。在利兹市可以找到空置的店铺去经营,但在约克市这根本不可能。房产异常珍贵,对于一个城市来说,房产适合旅游业的发展,至少适合备受瞩目的艺术品展出。当地许多居民在尝试投资规模更小但风险更高的项目,特别是为约克市当地居民提供服务的项目,这会是件完全不同以往的事情。"

　　希望使用本指南的读者下次前往约克市时,可为这些梦想家们提供一些支持。

音乐

克雷森特酒吧

　　作为先前的工人俱乐部和歌舞表演场地,克雷森特酒吧现正常营业,装配有Irie Vibes音响系统。尽管仍保留传统社区氛围、飞镖队和台球桌,但克雷森特酒吧现在已成为约克市重要的创造力发源地。本月,Josie Long、Mr Scruff、Acid Mothers Temple、Mark Thomas和Peggy Seeger等乐队在这个酒吧演出,不拘一格的形式使这里不仅有地下音乐和喜剧表演,还成为重点培养约克市本地人才的国际舞台。

　　克雷森特酒吧距离约克市中心仅几步之遥。无论是星期日街头食品活动(上一次活动在11月19日结束)还是漂亮的啤酒花园,都令这个酒吧在当地非常有名。最重要的是,克雷森特酒吧还是当地吉他乐队和各地巡演吉他乐队表演以及其他试验性活动的聚集场所。例如,该酒吧曾在11月22日主办EMOM York开放式电子音乐夜活动。"坏坏的查普尔(Bad Chapel)"本月在这里上演了Pye Corner Audio乐队的表演,其共同发起人彼得森声称:"克雷森特酒吧的经营者克里斯·谢林顿(Chris Sherrington)和克里斯·图克(Chris Tuke)对音乐非常认真,其中一位曾经是专业音响工程师。作为真正的音乐发烧友聚集场所,这个酒吧的设备让人惊叹。"

地下工作室

每个城市都需要一间灰暗的地下工作室。在低矮的拱形天花板下,约克市的俱乐部成员在这里忘情工作,沉迷于前沿的豪斯音乐、数码音乐和低音音乐中,直至凌晨时分。擅长使用 Mansion's Funktion-One 音响系统的 BlackBox 和 Animaux 发烧友将在 11 月 25 日参加"不计其数(Untold)"音乐活动。

费布尔斯

2014 年乔迁新居之后,这个流行音乐聚集地依然备受当地人欢迎。无论是这里经常举办中等规模的乐队巡演,还是偶尔举办悼念活动,都非常重视中年人的怀旧情怀(今年冬天,Shed Seven、Big Country 和 Zodiac Mindwarp 乐队曾在费布尔斯表演)。在费尔斯,除了 Lydia Lunch、纽约"brasshouse"三重唱和 Too Many Zooz 表演,这里还会举办其他前卫活动。

6.

【参考译文】

摩洛哥索伟拉市

去年"空巢"期,我们决定去享受冬日阳光。当然,我们希望费用越低越好。因此,我们选择从卢顿乘坐易捷航空的航班(票价仅 30 英镑)前往索伟拉。我们住在美丽的利雅得达尔阿杜尔酒店(Riad Dar Adul,每晚两人 49 英镑),从当地的美味鱼餐厅到卡斯巴艺术,我们把这个西海岸小镇麦地那的所有特色都体验了一遍。坐在无边无际的白色沙滩上享受 22℃的阳光,这次十一月旅行令我们终身难忘。

泰国利普岛

无论是如婴儿肌肤般柔软的白色沙滩、水晶般清澈的海水、新鲜的水果冰沙店,还是不限量的泰式炒河粉,都欢迎您到来体验!您只需缴纳两笔强制性参观费,即 1.20 英镑的租船费和 4.50 英镑的国家公园门票,即可乘船前往名为"利普"的那座小岛。我非常喜欢在 A-Plus 酒店居住,每晚收费仅 15 英镑,即可享用干净的住宿环境、强劲的空调和淋浴服务。在享用完酒店提供的免费新鲜水果之后,步行 200 米即可到达海边。在海边,您可享受日光浴,还可参加浮潜和跳水活动。当然,晚上轻啜泰国冰爽的"泰象"啤酒、享受安达曼海美景是度假的必备项目。

卡桑德拉·德尔霍

南非斯泰伦博斯市

利用汇率优势,您可以前往充满活力的大学城斯泰伦博斯市享受冬日阳光。斯泰伦博斯距离开普敦仅 1 小时车程。作为南非第二大古老城市,斯泰伦博斯坐落在开普酒乡。这里不仅有橡树荫蔽的街道、美妙的艺术和咖啡文化,还有当地形形色色热情的民众。我们居住在好齐尖酒店(HomeAway)。一月/二月期间,该酒店每晚住宿仅需 23 英镑。斯泰伦博斯是吃货天堂,市内遍布世界级的餐馆。在塔阿芙(Twaalf)社区餐馆,您不仅可以品尝到美味的烤肉(braai)和当地其他佳肴,还可以向无家可归的人们提供帮助。

特纳利夫岛洛斯克里斯蒂亚诺斯

请忘掉您已了解的特纳利夫南海岸信息,动身前往洛斯克里斯蒂亚诺斯。在这里,您可以享受多日灿烂的阳光、数不胜数的美景以及迷人的沙滩。在旅游胜地,游客可以自己去体验有趣的旅游项目,比如前往传统的"周末消遣"餐馆(Restaurant El Cine)。在"墙洞"("hole in the wall")小咖啡馆,您可以品尝一杯当地特有的"可塔朵"(cortado)和"塔帕"(tapa)咖啡。谁能抵挡在圣诞节漫步沙滩的诱惑呢?

乔迪·莱温图斯
印度果阿邦

我们在果阿角最南侧的果阿邦度过了一个美妙的假期,刚刚归来。虽然果阿邦的部分区域正遭受破坏,但这个位于印度南部的隐秘角落依然保留着传说中果阿当地古老的传统而极具魅力。我们居住的沙滩小屋坐落在私人沙滩上,从这里可尽情欣赏美丽的落日景色。此外,我们还租了轻便的助动车,前往帕洛琳(Palolem)享受奇妙的夜晚之旅,在雅高达(Agonda)和可拉(Cola)沙滩度过了慵懒的日子。尽管这不是印度最便宜的海边小屋(每晚两人140英镑;在雅高达每晚仅4英镑),但这绝对是印度最豪华的酒店。在我们居住的酒店中居然还配备有"极可意"水流按摩浴缸(jacuzzi)! 旅游时别忘了骑车去贝图尔堡(BetulFort)和萨尔(Sal)河口,穿过悠闲的村庄,体验葡萄牙殖民时期建成的古老房屋。这是一段美妙的旅程……

7.

【参考译文】

民主党态度转变对2020年美国总统选举影响重大

2017年9月22日,在旧金山召开的2017年度加州护士协会/国家护士组织委员会的卫生保健集会上,参议员伯尼·桑德斯(I-Vt.)发表讲话。桑德斯正在参议院推广他的"全民医保"法案。(图片由贾斯汀·沙利文/盖蒂提供)。

专栏作家:丹·巴尔兹(首席记者),10月7日,10:05

尽管在美国政治中党派分歧已司空见惯,但并不能就此断言共和党出现右倾化趋势是共和党两极分化加深的原因之一。近期,民主党内出现的左倾趋势同样值得关注,在民主党期望赢得2020年大选的背景下,这种影响更加明显。

以下是相关背景。2008年,民主党提名的所有主要候选人均不赞成将同性婚姻合法化,但到2016年时,民主党党内所有候选人无人反对同性婚姻合法化,这个态度的转变不只是因为最高法院的裁决。所有选民(特别是民主党选民)态度的变化使所有候选人相信,若想领导民主党,就必须支持同性婚姻。

贸易政策是另一个研究案例。多年来,民主党内部对多边自由贸易协定的优点存在分歧。1992年,在工会和其他组织强烈反对下,比尔·克林顿强烈支持北美自由贸易协议(NAFTA)。克林顿在工会大厅宣传其计划,尽管未能成功说服对手,但借助此次反对浪潮其开始在政坛上攀升。

2016年,随着对贸易和全球化的普遍怀疑,希拉里·克林顿不愿为自由贸易协定的优点提供类似的辩护。总统候选人参议员伯尼·桑德斯(I-Vt.)抨击《跨太平洋伙伴关系协定》(TPP),而希拉里加入了反对该协定的阵营。尽管在就任国务卿期间,希拉里曾对该协定的前景发表过热情洋溢的支持性评论,但最终她还是成了时任总统巴拉克·奥巴马的对手。

展望2020年,在卫生保健方面也可能会出现类似状况。在完成"单一付款人卫生保健

计划"方面存在诸多难题,但由于党内态度已发生变化,任何民主党人若不支持该计划将很难获得候选人提名。

在2016年总统竞选期间,桑德斯大力倡导其"全民医保"计划。然而,克林顿表现得更为谨慎,他认为桑德斯所倡导的"全民医保"计划存在缺陷,主张对《平价医疗法案》加以改进。

现在,桑德斯在参议院中推广"全民医保"措施,其共同发起人包括几位极可能在2020年获得民主党提名的总统候选人。

与此同时,为进一步推进共和党人约翰·科尼尔斯(D-Mich.)发起的"单一付款人方案",绝大多数国会民主党议员均已加入该计划。尽管如此,赞成科尼尔斯想法的议员原则上也同样质疑现在是否为实施该计划的黄金时间,这不仅是因为该计划存在潜在成本和缺乏适当支付机制问题,还因为该计划也存在其他潜在的政策缺陷。

促使民主党人支持"单一付款人方案"的压力来源于民主党内部态度的转变。6月,皮尤研究中心发现,当前52%自称是民主党人士的民众支持政府实施的卫生保健体系。与今年年初相比,这一数字已上升了9个百分点,自2014年以来上升了19个百分点。在自由民主党人中,64%支持该计划(仅今年就已上升了13个百分点),而在年轻的民主党人中,66%支持该计划。

民主党态度显著转变的领域除了卫生保健,还包括政府职能、社会安全网络、社会中的种族角色和种族歧视、移民,以及多样性价值等领域。

几天前,皮尤中心针对共和党和民主党之间日益扩大的分歧实施了一次全面调查,调查结果的核心内容可以用该中心在研究报告开篇中的一句话总结:"共和党和民主党在思想上的差异达到过去20多年的顶峰。"

这次民意调查是自1994年以来进行的一系列调查中最新的一次。这些调查不仅及时提供当前最新信息,还表明公众舆论变化过程。共和党右倾的时间比民主党开始左倾的时间更早,程度更大,且共和党的政策立场更加坚定。但是,就一系列问题上的态度而言,共和党的转变幅度最大。

在最新实施的调查中,皮尤发现党派分歧最大的焦点在于政府是否应帮助困难民众,这主要是因为民主党人最近态度上的转变。从2011年到现在,认为政府应该向那些需要帮助的民众提供更多帮助的民主党人比例已从54%上升到71%。

仅少数(24%)共和党人认为政府应该为那些需要帮助的民众提供更多帮助,而该比例在过去六年中几乎没有发生改变。对于这一问题,共和党人的态度曾在2007年至2011年期间(即在奥巴马就任总统的最初几年)发生转变。在这期间,共和党人对政府角色的支持率下降了20个百分点。

民主党人对上述两个相关问题的态度发生过类似转变。四分之三的民主党人认为"穷人生活艰苦是因为政府提供的福利远不足于帮助他们过上体面生活。"在过去几年,持该观点的人所占比例上升了12个百分点。

80%的民主党人认为国家需要继续变革,使黑人享有与白人平等的权利。自2014年以来,该比例上升18%。超过60%的民主党人声称:"种族歧视是大多数黑人如今无法获得成功的主要原因"。而3年前,该比例仅为40%。

与此同时,仅有四分之一的共和党人同意政府应提供更多的福利,不到40%的共和党人认为国家需要继续努力为黑人提供平等权利,仅14%的共和党人认为种族歧视是大多数黑人无法取得成功的主要原因。

近年来,两党成员对移民的态度均变得更为乐观,但最新的调查结果显示,两党对此仍存在42%的巨大差距。现在,84%的民主党人认为移民通过努力工作和个人才华增强了美国实力。2010年,持此观点的民主党人士比例为48%,而共和党人的比例为29%,现在共和党人的支持比例已升至42%。

为何民主党的立场发生如此大的变化?为何民主党近期在种族、政府和移民问题上态度发生转变?尽管调查报告对此没有给出明确解释,但一个原因可能是这与2016年总统竞选活动和特朗普当选总统有关。

无论是呼吁在美墨边界修建隔离墙,还是禁止主要来自中东国家的难民进入美国,特朗普总统都坚定支持他自己充满争议的移民立场。此类立场帮助特朗普赢得总统宝座,但相应政策和特朗普经常给出的超前言辞也引起总统反对者的强烈抨击。

2016年总统竞选活动的结束突显美国存在国家认同(种族、移民、国家性格和社会面貌变化方面)危机,8月的夏洛茨维尔骚乱突显了这种丑陋的分歧。

特朗普当选总统以及普通民主党人对总统的反应,令人们对民主党有了新的认识。自从特朗普宣誓就任总统以来,民主党领导人的态度一直备受关注并且不断变化。

同样受到上述情绪影响的还有期望在2020年获得党内总统候选提名的民主党人。他们必须通过适当的方式,利用民主党态度的变化提出一个吸引民主党之外选民的政治愿景,这一政治愿景不仅以反对特朗普的怨恨情绪为基础,更应建立在鲜明健全的政策之上。但是,在民主党和共和党态度两极分化时期,要做到这一点实属不易。

8.

【参考译文】

股价暴跌之后美联航首席执行官温和道歉

针对周日一名男性乘客遭暴力驱逐下机事件,美联航首席执行官再次公开道歉,称这一事件"真的很可怕。"

周二,美联航首席执行官奥斯卡·穆尼奥斯(Oscar Munoz)于该视频(乘警强行将该男子拖出机舱)被同机乘客发布一天之后,在向员工发送的信函中写道:"任何人都不应遭受这样的待遇"。

周一,穆尼奥斯在官方声明中将暴力驱逐事件轻描淡写为对乘客的"重新安置",称这位男性乘客的行为具有"破坏性和好斗性"。紧接着,穆尼奥斯受到民众谴责。周二,美联航股价暴跌,穆尼奥斯却把注意力重新转回公司。

在周二的股票交易中,美联航的市值蒸发近10亿美元。当天晚些时候,穆尼奥斯发表声明表示,他决定"纠正错误,保证类似事件不再发生。"他承诺将审查美联航在"鼓励乘客自愿放弃乘机、处理机票超售航班,以及与机场管理当局和当地执法部门合作"等方面的政策。

美联航计划在4月30日之前向公众发布审查结果。

美联航母公司——联合大陆控股公司的股价周一中午之前暴跌4%,市值蒸发近10亿美元,随后小幅上涨,截至周一收盘时仍下跌2.8%,较周一收盘时的市值225亿美元,蒸发约 6 亿美元。

在周一的交易中,投资者基本未受到美联航所处困境的影响。该航空公司股价在周一收盘时候上涨 0.9%,公司市值增加约 2 亿美元。

上周日,在芝加哥奥黑尔国际机场,一名男性乘客拒绝在美联航超售航班上自愿放弃乘机,乘警暴力将其驱逐下机。自周日以来,美联航的问题似乎因此事件而持续发酵。

一段视频显示,乘警强行抓住一名乘客,然后暴力拖拽到客机过道上,其他乘客惊呼"我的天哪""看看你们对他做了什么"。

白宫新闻秘书肖恩·斯派塞(Sean Spicer)声称,他"确信"唐纳德·特朗普(Donald Trump)总统已看过这段视频。

斯派塞对记者称:"我认为任何人在看到那段视频后都无法保持冷静,任何人都不应遭受这种对待。在视频中我们可清楚地看到这名乘客在客机走道上被拖着走,他的脸因撞到扶手或其他什么地方而鲜血直流。我认为在看到这种场景时,任何人都不会无动于衷。当讨论另一个人遭受的这种不公平待遇,你无法仅轻描淡述的说我们会处理好这件事。"

斯派塞指出,美联航正在调查这一事件。他认为该事件是地方事件,无需联邦政府回应。"我认为多家执法机构均有权审查这一事件。我知道美联航已表示正在审查公司政策,在审查结果出来之前,我们无需过度猜测。"

周二,穆尼奥斯因称这名乘客具有"破坏性和好斗性"并称机上员工的行为"符合规定程序"而遭受民众谴责。

在视频中可以听到该乘客自称华裔居民。该视频在中国已引起民众一片愤慨,中国民众呼吁抵制美联航。

股票市场开盘之前,美联航股价在盘前交易中已下跌 6%。

9.

【参考译文】

远程医疗在主流卫生保健领域有很大的发展前景。现在业内针对患有心脏病或糖尿病等疾病的患者正在实施大量试验,以评估技术。待技术成熟时,即使患者待在舒适的家中,也可接受监测。已植入起搏器的患者现在通过手机即可检查心脏起搏器运行状况,无需定期就诊检查。这样,患者仅需要在绝对必要时前往诊所就诊即可。

同样,德国 BodyTel 公司是已开发出传感器的几家公司之一。该传感器利用蓝牙无线技术,可测量患者血糖水平、血压水平和体重,并将测量数据上传至安全网络服务器。患者利用该技术可管理和监测自己的身体状况,向医生提供不断更新的信息。美国工业巨头企业霍尼韦尔公司已设计出一套系统。利用该系统,患者可在家自行检测肺部功能、心电图、血氧饱和度和血压,以监测肺病、充血性心力衰竭等病况。医生可查看评估数据。医生若发现患者身体有异常状况,可采取比如更换药物等适当措施对患者进行救治。

远程医疗技术对于患者和医护人员均具有强烈的吸引力。利用此类技术,患者可独立照顾自己,在家中生活,成本也比在医院治疗低。随着发达国家逐步进入老龄化社会,患者不但可在家接受治疗,而且还可以省钱,远程医疗技术应用前景愈加具有吸引力。

当前不只是已确诊患者开始接受使用这种设备。自 2006 年起,英国已向 16 万名老年居民提供专用设备,以便老人在家居住,在"预防性技术补助"方面已花费 8 000 万英镑(约合 1.6 亿美元)。

然而,现在的大多数技术需要患者牢记自我监测并操作相应设备。对于部分患者(如处于阿尔兹海默症早期阶段的患者)来说,此类要求可能不切实际。因此,为实现监测过程自动化以及设备更方便使用,许多工作正在进行中。

比如,加利福尼亚大学洛杉矶分校的威廉姆·凯斯勒(William Kaiser)与其同事现已开发出为康复患者提供监控和建议的"智能手杖"。凯斯勒博士声称:"该智能手杖配有可测量手杖顶端和把手周围压力的力传感器、运动传感器和加速度计。"这样,智能手杖可通过计算患者步态而得知患者使用手杖的方式,向患者(如髋关节置换手术患者)提供更好地使用手杖的反馈建议。凯斯勒博士说:"智能手杖可通过蜂鸣声或仿真声音向患者提供指导性建议。"

另一种应用是使用家庭嵌入式传感器。瑞士房屋管理企业 Oliver Goh of Implenia 公司现已开发出一套可检测房屋居住者健康状况的系统。利用安装在门和床垫上的传感器,按照设置自动打开的智能药盒、心脏检测器和位置感应腕表,该系统可帮助护理人员监测在家居住的老人。高(Goh)先生声称:"该公司已招募了六名愿意接受技术测试的老年志愿者。如果老年人不幸心脏病发作,或无法起床,或需要他人帮助,我希望利用该技术护理人员很快就发现此类状况。从根本上讲,我们想知道这种方法是否会有助于老年人延长预期寿命。"

预防比治疗更为重要。

进一步看,远程医疗技术可能会对"走路正常"的健康人群有吸引力。如果传感器可在不侵犯使用者隐私或不影响使用者舒适生活的条件下完成监测,医生就能在患者出现任何病症之前诊断出疾病。IBM 公司的格兰迪博士声称:"该技术正从远程医疗领域转移至远程健康和远程预防领域。"他补充道,该技术可进一步提高卫生保健系统的效率。

凯斯勒博士声称,这种医疗方法不仅可减少费用,也可在早期发现疾病。他说:"医生可在治疗成本较低、对患者影响较小的早期阶段检测到疾病。"美国远程医疗协会的常务董事约翰·林克斯(John Linkous)承认,这种技术目前还不存在。他说:"尽管目前还做不到用一台设备即可完成全身检查,但从长远来看,我对该设备的出现非常乐观。"

一种开发思想是使用无线红外皮肤传感器测量血球计数、心率和血氧水平,另一种开发思想是向人体植入可由身体热量供电的无线传感器。还有一种更普遍的开发思想是,通过使用智能马桶监测人类粪便,来查看人们是否患有肠道疾病或癌症迹象。筛选结果时,林克斯博士声称,该技术最难的并非开发传感器技术。他说:"该技术应用时会产生海量数据,问题是医疗卫生系统尚无法处理这些海量的数据。"

巴加拉(Bakalar)医生声称,要解决这一难题就需要开发更多的技术。他说:"我们必须想办法过滤信息,以避免医疗服务系统信息量大。"最明显的方法是使用"专家系统",这是利用专家医疗知识、提供临床诊断结果的软件。

和远程医疗技术类似,"专家系统"已存在一段时间了。在丹麦开展的试验中,已利用"专家系统"软件向医生提供处方建议,这表明该软件拥有广泛的应用前景。有时,"专家系统"提供的临床诊断结果甚至比医生更准确。当前,该软件并没得到广泛应用,部分是因为医生不愿意受放在角落的计算机摆布,其他原因还包括该软件当前仍无法融入临床医疗。

然而,远程医疗技术可能是远程健康领域中的理想技术。通过在后台筛选传感器产生的数据,远程医疗技术仅在必要时向医生发送警报或提醒。

远程医疗技术向远程健康技术的转变说明该技术可从诊断和治疗领域扩展至"保持健康"这一更为广泛的领域。根据技术开发结论可见,远程健康技术将使用无线传感器和人体植入装置,对所有人群实施筛选,在人们的日常生活中发现疾病早期迹象。广泛普及远程健康技术之后,感到不舒服而预约医生的日子将一去不复返。取而代之的是医生向您致电,预约治疗疾病。

10.

【参考译文】

每年英格兰约10万家庭无购房能力

一份报告显示,因缺少民众可承受的租赁性或购买性住房,每年英格兰增加30 000户无住房家庭。

秋季的声明重申,政府计划每年出资建设300 000套住房。摄影:丹妮拉·哈瑞拉/瑞雷克斯/苏特·斯托克

一份报告显示,因缺少可租赁或可购买的房屋,每年英格兰约10万户家庭无法获得住房。

房地产代理机构第一太平戴维斯公司(Savills)与《卫报》联合开展的一项研究表明,本年度英格兰买不起高价房的家庭数量已比之前预计的2015年的70 000户有所增加,部分原因是房屋需求评估模式发生改变,还有房价上涨和工资增长停滞等原因。

周三,英国财务大臣菲利普·哈蒙德(Philip Hammond)再次重申,英国政府承诺到21世纪20年代中期,每年出资建设300 000套住房,同时公布了一系列支持住房建设的新措施。第一太平戴维斯公司指出,为了满足英国民众对保障性住房日益增长的需求,三分之一的政府建设住房价格应低于市场价格。

该公司声称,按照市场利率计算,约有96 000户家庭不具有购买和租赁住房的能力,绝大部分这样的家庭位于伦敦市和英国南部地区。该公司还指出,英国政府应采取多种措施解决国内不同地区的房屋短缺问题:对于收入较低的市场区域,需要更多低成本租赁住房;对于房价较高的市场区域,适合实施包括共有产权在内的混合型住房。

该公司声称,在伦敦市内,20%受影响的家庭年收入超过35 000英镑;同时还有20%的家庭年收入不足10 000英镑。

该研究发现,在过去三年内,英国每年新建的55 000套保障性住房低于民众需求数量。尽管伦敦市每年有42 500户家庭需要低于市场价格的住房,但每年政府能交付使用的保障性住房仅有8 800套。在伦敦市外以南区域,每年政府新建约15 500套保障性住房,但每年民众需求量却高达34 100套。

同时,在英格兰北部区域,低收入家庭每年对保障性住房的需求量为9 600套,然而政府每年可交付使用的保障性住房为8 900套。

英国政府年度预算案公布后,新闻头条大幅宣传哈蒙德(Hammond)取消大多数首次购房者印花税的决定,但这位财务大臣指出,政府将提供更多资金建设保障性住房,改变房地产市场现有规则,以帮助地方议会提供住房。

  今年10月,英国首相特蕾莎·梅在保守党会议上承诺,未来四年英国政府将提供20亿英镑资金支持社会性住房建设。然而,第一太平戴维斯公司认为,英国政府每年需要提供70亿英镑资金,才能在四年内彻底解决10万户家庭的住房问题。

  为新建保障性住房提供资金,可能会使英国政府的房地产市场税收每年减少4.3亿英镑。

  第一太平戴维斯公司住房顾问部门主管海伦·柯林斯(Helen Collins)声称,英国政府公布的预算将为英国房地产市场带来诸多积极因素,但他也表达了一些担忧。

  柯林斯声称:"尽管财务大臣带来大量利好消息,我们仍感觉到住房政策依然存在一些特别重要的问题尚未解决。"

  此类问题包括英国政府并未改变在绿化带用地上建房的做法,也未承诺向社会化租赁房屋提供更多资金。柯森斯还指出:"对低收入家庭来说,社会化租赁住房比私人租赁住房更好、更便宜,这一点却几乎无人提及。"

二、汉译英

1.

【参考译文】

Euphorbiasplendens Bojer ex Hook.

(Source: Southern China Botanic Garden; Date: October, 23, 2014)

  Euphorbia splendens Bojer ex Hook. is also called Crown of Thorns, Euphorbia Milii, or Panzai Thorn. It is classified as a deciduous shrub in the Euphorbiaceae family and is native to the African country of Madagascar and its archipelago. Its flowers, which appear in pairs at the ends of long stalks and whose stems are densely covered with hard brown thorns, are tiny and without petals. The bright red kidney-shaped parts are bracts that are often taken for petals because of their vivid color. The flowering season for this plant is quite long, extending from March to December. The blossoms are at their most beautiful during the spring season. Because they look like wintersweet flowers with stiff thorns on the stem, the plant is called Christ plant/Crown of Thorns" (Euphorbia Milii). Having a comparatively long life span, the potted plant can live for decades. Moreover, owing to its slow growth, it can be tied into a variety of shapes to form a mini-landscape. Under bright sunlight, its flowers are dazzling (dazzle the eyes). Because it does not wilt and is long-lasting, it is the perfect potted plant for indoor decoration.

2.

【参考译文】

  Abstract: This paper is an introduction to *the English-Chinese Translation Course Book* published by Higher Education Press. Using the framework of cohesion theory, we explain our notion that in translation theory guides practice while practice both confirms and enriches theory. This paper discusses four aspects of grammatical cohesion: reference, substitution, omission and conjunction. The examples cited in this paper are taken from the above mentioned textbook.

3.

【参考译文】

Opinions on Further Perfecting the Pharmaceutical Production Supervision Mechanism of Guizhou Province in 2014

In order to further perfect the pharmaceutical production supervision and management mechanism, define working responsibilities, improve working efficiency and facilitate enterprise services, the following working opinions on perfecting the effect of the pharmaceutical production supervision working mechanism on research decisions have been put forward in the spirit of the following: *the Decision of the State Council on Matters concerning a Group of Administrative Approval Items to Be Cancelled and Delegated to Lower Levels* (GF〔2013〕No. 19), *Regulation on the Administration of Narcotic Drugs and Psychotropic Drugs* (Order No. 442 of the State Council), *Measures for Supervising and Administering Pharmaceutical Production* (No. 14 of 2004 Order of the Food and Drug Administration) and *the Notice on Printing and Distributing the Work Points of the Chinese Food and Drug Administration in* 2014 (SYJ〔2014〕No. 7).

Ⅰ. Before the preparation of the declaration, the pharmaceutical production enterprise shall first submit it to the food and drug administration either at the municipal or provincial level for the first review and on-site checking. The administration will then provide the review with comments after the enterprise meets relevant local government industrial development planning and declaration conditions. The enterprise will submit them to the food and drug administration at the provincial level for approval.

Ⅱ. When the pharmaceutical production enterprise applies for GMP authentication, the food and drug administration at the municipal or provincial level shall carry out the first review and on-site checking after which they will present their comments. The enterprise will then submit them to the food and drug administration at the provincial level for approval.

Ⅲ. If the quality director and production director of the pharmaceutical production enterprise are changed, the enterprise shall submit the resume, academic certificate and other related documents of the changed personnel to the food and drug administration at the municipal or provincial level for registration within 10 working days after change. The food and drug administration at the municipal or provincial level shall then fax these documents to the food and drug administration at the provincial level for the record within 5 working days.

Ⅳ. If the quality authorizer of the pharmaceutical production enterprise is changed, theenterprise shall submit the resume, academic certificate and other related documents of the changed personnel to the food and drug administration at the municipal or provincial level for the first review, and then submit this information to the food and drug administration at the provincial level for filing.

Ⅴ. If the key production facilities and conditions and situations at the pharmaceutical production enterprise change, the enterprise shall report to the food and drug administration at the municipal or provincial level for registration within 20 working days after the change and report to the food and drug administration at the provincial level for a review within 10 working days. The food and drug administration at the provincial level may then entrust the food and drug administration at the municipal or provincial level with checking the situation according to the need and filing this case meeting the requirements of Good Manufacture Practice of Medical Products.

Ⅵ. For narcotic drugs, psychotropic drugs and pharmaceutical precursor chemicals, the pharmaceutical production enterprise engaged in the production of ordinary preparations and the designated production of psychoactive Ⅱ shall report the purchasing, application and sales of these special drugs to the food and drug administration at the municipal (state) level in writing every month. The food and drug administration at the municipal (state) level shall report conditions relating to special drug supervision to the food and drug administration at the provincial level every 3 months, and implement the system of reporting special drug supervision to the leaders of the administration on a quarterly basis.

Ⅶ. If the supplier of the pharmaceutical production enterprise changes, the enterprise shall, for the record, report the change to the food and drug administration at the municipal or provincial level within 15 working days after the change.

Ⅷ. If the drugs are commissioned for production and extraction & processing, the commissioning party of the pharmaceutical production enterprise shall report to the food and drug administration at the municipal level or state level for the first review and then to the food and drug administration at the provincial level for approval. The local food and drug administration of the commissioning party is responsible for the daily supervision of production.

Ⅸ. The above working mechanism will be refined before it is implemented from Apr. 2014.

4.

【参考译文】

A Letter of Recommendation

The undergraduate student, ×××, who is majoring this year (2006) in Finance at School of Economics, Sichuan University is a student in my ××× class. This student is positive, conscientious, and pragmatic, gets good grades, and ranks top in the class.

When teaching this course, I focus on developing students' ability to use basic theory and basic knowledge to analyze and think about problems, as well as on their ability to calculate and perform operations. This student has performed outstandingly during the time s/he has been in my class. S/he not only listens attentively in class and puts forward original questions, but also has the courage to take the floor and clearly express his/her point of view regarding my case analysis questions. This student is a real thinker who actively reflects on problems and probes deeply intotheir essence while demonstrating a marked ability for problem analysis. This student has a solid professional foundation and is flexible as to the application of basic theory and knowledge.

　　Based on his/her outstanding performance, I strongly recommend the above named student and hope that he/she will be accepted.

Recommended by: ×××

参 考 文 献

[1] CHEN Q J. The Application of Text Type in Non-literary Translation Teaching [J]. Translation & Interpreting Studies,2010,5(2):208-219.

[2] FLÖCK I,GELUYKENS R. Speech Acts in Corpus Pragmatics:A Quantitative Contrastive Study of Directives in Spontaneous and Elicited Discourse[C]//TRILLO J R. Yearbook of Corpus Linguistics and Pragmatics 2015:Current Approaches to Discourse and Translation Studies. Switzerland:Springer International Publishing, 2015:7-37.

[3] MÁRTA L. Conceptualizing Translation Competence [J]. Translation & Interpreting Studies,2010,5(2):167-194.

[4] PLAZA S M. Black and White Metaphors and Metonymies in English and Spanish: A Cross-Cultural and Corpus Comparison[C]//TRILLO J R. Yearbook of Corpus Linguistics and Pragmatics 2015:Current Approaches to Discourse and Translation Studies. Switzerland:Springer International Publishing,2015:39-63.

[5] WANIEK-KLIMCZAK E. Aspiration in Polish:A Sound Change in Progress? [C]//PAWLAK M,BIELAK J. New Perspectives in Language,Discourse and Translation Studies. Berlin Heidelberg:Springer,2011:3-11.

[6] WYKE B V. Exploring the Author-Translator Dynamic in Translation Workshops [J]. The Translator,2012,18(1):77-100.

[7] 陈刚. 旅游翻译[M]. 杭州:浙江大学出版社,2014.

[8] 程同春,程欣. 现代科技英语阅读教程[M]. 南京:东南大学出版社,2013.

[9] 丛晓峰,余刚,谢斌,等. 基于模糊数学法的新建绿地土壤物理性能评价[J]. 中国农学通报,2018,34(29):59-63.

[10] 范莹芳. 新编科技英语阅读教程[M]. 哈尔滨:哈尔滨工业大学出版社,2011.

[11] 范祥涛. 研究生科技语篇英汉翻译教程[M]. 苏州:苏州大学出版社,2011.

[12] 方梦之,范武邱. 科技翻译教程[M]. 上海:上海外语教育出版社,2015.

[13] 方梦之,傅敬民. 振兴科学翻译史的研究:应用翻译研究有待拓展的领域[J]. 外国语(上海外国语大学学报),2018(3):67-75.

[14] 方梦之. 英语科技文体:范式与翻译[M]. 北京:国防工业出版社,2011.

[15] 冯志杰. 汉英科技翻译指要[M]. 北京:中国对外翻译出版公司,1998.

[16] 冯志杰. 中国近代翻译史:晚清卷[M]. 北京:九州出版社,2011.

[17] 傅勇林,唐跃勤. 科技翻译[M]. 北京:外语教学与研究出版社,2012.

[18] 郭立秋,范守义,贾令仪. 语篇翻译中的衔接问题:理论解读与翻译应用[J]. 上海翻译,2011(4):31-34.

[19] 韩孟奇. 科技英语阅读[M]. 上海:上海交通大学出版社,2012.

[20] 胡静. 航空器制造工程专业英语[M]. 西安:西安电子科技大学出版社,2015.

[21] 姜华. 计算机媒介语篇中的词汇特点[D]. 大连:大连理工大学,2007.

[22] 居祖纯. 汉英语篇翻译[M]. 北京:清华大学出版社,2000.

[23] 康志洪. 科技翻译[M]. 北京:外语教学与研究出版社,2012.

[24] 李健. 科技英语阅读教程[M]. 北京:外语教学与研究出版社,2015.
[25] 黎淑翎,周剑云. 快速公交车站建筑气候适应的主要策略初探[J]. 华中建筑,2018(11):124-128.
[26] 刘瑜. 民主转型与政治暴力冲突的起落:以印尼为例[J]. 学海,2017(2):45-55.
[27] 刘志武,杨琼,唐宝昌. 航空专业英语[M]. 北京:北京理工大学出版社,2015.
[28] 马会娟. 汉译英翻译能力研究[M]. 北京:北京师范大学出版社,2013.
[29] 马祖毅. 中国翻译简史:"五四"以前部分[M]. 增订版. 北京:中国对外翻译出版公司,1998.
[30] 浦建君,陈艾莎. 民航专业英语[M]. 成都:西南交通大学出版社,2012.
[31] 钱三强. 医学名词:第一分册[M]. 北京:科学出版社,1989.
[32] 秦荻辉. 精选科技英语阅读教程[M]. 西安:西安电子科技大学出版社,2008.
[33] 沈星. 航空科技英语[M]. 北京:北京理工大学出版社,2015.
[34] 孙昌坤. 实用科技英语翻译[M]. 北京:对外经济贸易大学出版社,2013.
[35] 孙先洪. 高校教师计算机自我效能感与计算机态度的实证研究[J]. 江苏教育,2017(5):64-66.
[36] 谭卫国. 英语新闻的用词特点初探[J]. 外语与外语教学,2004(2):11-14.
[37] 谭载喜. 西方翻译简史[M]. 北京:商务印书馆,2000.
[38] 陶涛. 民航英语特点及翻译策略浅谈[J]. 中国民航学院学报,2012(6):13-16.
[39] 田建国. 英汉航空航天科技大词典[M]. 西安:西北工业大学出版社,2018.
[40] 田娟,陈斯,胡杨. 航空英语[M]. 天津:天津大学出版社,2017.
[41] 汪洪梅. 新闻语篇的翻译教学研究[J]. 西部素质教育,2017,3(22):182.
[42] 王银泉. 实用汉英电视新闻翻译[M]. 武汉:武汉大学出版社,2009.
[43] 王燕. 医学英语翻译与写作教程[M]. 重庆:重庆大学出版社,2008.
[44] 吴绍健,林维成. 英汉·汉英科技搭配词典[M]. 北京:化学工业出版社,2006.
[45] 夏廷德,马志波. 实用新闻英语翻译(英汉双向)[M]. 北京:对外经济贸易大学出版社,2010.
[46] 许建平. 英汉互译实践与技巧[M]. 4版. 北京:清华大学出版社,2012.
[47] 严俊仁. 汉英科技翻译新说[M]. 北京:国防工业出版社,2010.
[48] 严俊仁. 新汉英科技翻译[M]. 北京:国防工业出版社,2010.
[49] 闫文培. 实用科技英语翻译要义[M]. 北京:科学出版社,2008.
[50] 苑春明,姜丽. 商务英语翻译[M]. 北京:外语教学与研究出版社,2013.
[51] 张干周,郭社森. 科技英语翻译[M]. 杭州:浙江大学出版社,2015.
[52] 张乾友. 集体决策的政治与协议的政治:解决冲突的两种公平方式[J]. 行政论坛,2017(2):12-19.
[53] 张曦. 科技英语翻译教程[M]. 上海:上海交通大学出版社,2016.
[54] 张先刚. 评价理论对语篇翻译的启示[J]. 外语教学,2007,28(6):33-36.
[55] 赵萱,郑仰成. 科技英语翻译[M]. 北京:外语教学与研究出版社,2016.
[56] 中国人民解放军总装备部军事训练教材编辑工作委员会. 科技英语翻译实用教程[M]. 北京:国防工业出版社,2009.
[57] 周怡青,肖友利. 活性天然产物靶标蛋白的鉴定[J]. 化学学报,2018,76(3):177-189.